AF413233

SNAKE MEN

ALSO BY ZACH ST. GEORGE

The Journeys of Trees:
A Story About Forests, People, and the Future

SNAKE MEN

REBELS, REPTILES,
AND THE RACE TO NAME
EARTH'S CREATURES

ZACH ST. GEORGE

W. W. NORTON & COMPANY

Independent Publishers Since 1923

Copyright © 2026 by Zachary St. George

All rights reserved
Printed in the United States of America
First Edition

No portion of this work is authorized for reproduction, storage in a retrieval
system, or transmission by any means in any form, and all text, images,
and data herein contained are expressly excluded from use in any machine
learning, artificial intelligence training, or data mining activity, absent prior
written permission. The publisher waives no rights under domestic or international law and expressly reserves this work from the text and data mining
exception pursuant to Article 4(3) of the European Union Digital Single
Market Directive 2019\790.

For information about permission to reproduce selections from
this book, write to Permissions, W. W. Norton & Company, Inc.,
500 Fifth Avenue, New York, NY 10110

For information about special discounts for bulk purchases, please contact
W. W. Norton Special Sales at specialsales@wwnorton.com or 800-233-4830

Manufacturing by Lake Book Manufacturing
Book design by Daniel Lagin
Production manager: Lauren Abbate

ISBN 978-1-324-02168-1

W. W. Norton & Company, Inc., 500 Fifth Avenue, New York, NY 10110
www.wwnorton.com

W. W. Norton & Company Ltd., 15 Carlisle Street, London W1D 3BS

Authorized EU representative: EAS, Mustamäe tee 50, 10621 Tallinn, Estonia

1 2 3 4 5 6 7 8 9 0

For Tia. I don't know any new snakes,

but if I did, I'd name them all after you.

CONTENTS

SNAKE MEN

ADAM'S TASK

> And here let those
> Who boast in mortal things, and wond'ring tell
> Of Babel, and the works of Memphian kings,
> Learn how their greatest monuments of fame,
> And strength and art are easily outdone
> By Spirits reprobate, and in an hour
> What in an age they with incessant toil
> And hands innumerable scarce perform.
>
> —MILTON, *PARADISE LOST*

Late in the afternoon of a day in which Raymond Hoser had thrilled a crowd of children with his acts of skill and bravery, in which he had rendered aid to a number of imperiled citizens, some of them more imperiled than others, and in which he had driven his Nissan Micra across what seemed like most of the Greater Melbourne metropolitan area, he paused in cleaning yet more snake shit from the inside of yet another clear plastic tub. He was standing at the sink. With his right hand he was swabbing the tub. With his left hand he was holding the two snakes that had shit in the tub. They were eastern brown snakes, known to be among the deadliest types of venomous snakes in Australia, and therefore among the deadliest types of venomous snakes in the world. He was holding them with roughly the care one might take to hold a garden hose.

Hoser had spent much of the day describing the depravities and depredations of his enemies, who are legion and tireless, and who are led, he says, by the German snake-venom expert Wolfgang Wüster. So great were Wüster's scandals that, in Hoser's haste to recount them, he had at certain points missed important turns, forcing him to perform additional acts of skill and bravery in his Nissan Micra in order to make his appointments. Now he paused momentarily in both the cleaning and the retelling. He crossed the room, still clutching the snakes. He pulled down a title from the bookcase and brought it to the corner of his desk, next to where I was sitting. The book was called *Snakes of Western Australia*. He flipped straight to the page he wanted and pointed at a snake. I recognized the snake he was pointing at as being of critical importance, central to the whole story, but I found that at that moment, my attention was divided between, on the one side, that important snake from Western Australia, and on the other side, two important snakes from eastern Australia, whose heads were now bobbing several inches from my bare calf.

Raymond Hoser is the Snakeman. It is a title he has trademarked and at times jealously defended. He has also trademarked "reptile man," "man with snakes," "snake handler," "hold the animals," and a host of other phrases, including his own name. As a day job, Hoser runs Snakebusters, a Melbourne company that offers several services. One service is snake removal. This is a surprisingly common trade in Australia, akin to plumbing, with which it sometimes intersects; Hoser has removed numerous snakes from toilet bowls and other fixtures. Another service is snake shows, most often performed at children's birthday parties. Additionally, Hoser edits the *Australasian Journal of Herpetology*, a scientific journal, and has written one book about Australia's reptiles and frogs, one book about Australia's endangered animals, two books about driving a taxi, and five books about corruption and animal smuggling among Australian police and wildlife officials. These feats and accomplishments, along with Hoser's penchant for media attention and civil litigation, long ago brought him a modest, local kind of fame. But by the time I first heard of him, his fame had spread around the world. He had become known as a taxonomist, a namer of new species—and one of the most prolific to ever live. Over the preceding decades, he had named hundreds of new snakes, hundreds of lizards, dozens of turtles, four crocodiles, a hundred frogs, a couple dozen mammals, and a handful of fish. In the process, he'd set off an international imbroglio concerning, by some accounts, no less than the future course of life on Earth.

~

Biological taxonomy, which is the scientific study of the relationships among different kinds of living things and the process of describing, naming, and classifying them accordingly, is widely known to be among the most boring branches of science. It is the stultifying realm of solitary, formalin-scented obsessives who spend their days peering through microscopes, measuring subcaudal and rostral scales, muttering to themselves words like "cladistics," and "systematics," and "monophyly." It is a science of sifting, of sorting, of grouping like with like, of drawing lines between this and that, of plodding, methodical, repetitious labor. This utter dullness extends even to the name, "taxonomy," coined in 1813 by Swiss botanist A. P. de Candolle, in reference to the classification of living things. Deriving from the Greek *taxis*, meaning "arrangement," and *nomos*, meaning "method," the word has expanded to encompass all kinds of nested organizational schemas and structures, including in computer science, corporate hierarchies, and the Dewey decimal system—an expansion that has necessitated the addition of an even duller heading, "biological," to denote the type of taxonomy in question. Then there are the taxonomists' most public product, the two-part names they give creatures—Latinized, italicized, unpronounceable, unintelligible, impossible to remember. To linger too long on the subject of biological taxonomy will dry the mouth and shrivel the eyeballs in their sockets.

This, at least, is what I thought, to the degree I thought about biological taxonomy at all, which happened mostly when, in my role as a science reporter, one of the scientists I interviewed used the two-part scientific name for the creature they were discussing. These unpronounceable, forgettable names could usually be translated back into the common tongue; the broader science, meanwhile, could be mostly elided, sparing my readers the absolute drudgery of learning about it.

But I was wrong about taxonomy. Or, at least, my impression was incomplete. While I was correct on the broad details—about the formalin-scented obsessives, about their love of jargon and minutiae—I was missing a crucial, central part. It was like imagining skydiving to be mostly about folding parachutes or thinking that *Moby Dick* was the story of a weird white whale. I was oblivious to the whole soul of the matter.

My shadow-on-a-cave-wall view of taxonomy began to flicker when I noticed that, over the course of a several-years-long reporting project, when I tagged along with scientists on field trips, they often seemed unsure about the true identities of some of the creatures we encountered. This surprised me every time, since it seemed like such a basic piece of information. It was especially surprising since it was happening in places like Connecticut. Even if I didn't know the name of a given creature, surely the professionals I was with did; if they didn't, surely *somebody* did. But over and over, they said that they did not, and maybe, neither did anyone else.

Then, the precipitating event: One of those scientists put me in touch with a retired logger in coastal northern California's Del Norte County who thought he'd discovered a new species of redwood tree. It happened in the 1980s, in the last dying days of the timber industry in that part of the state. He worked as a scaler, judging the value of old-growth redwood logs that his colleagues dragged down from the hills. One day, a sawyer who was milling the logs into lumber told him that somehow red cedars had gotten mixed in with the redwoods. He went to see the logs in question. Based on his long experience, the wood, bark, leaves, and cones looked like redwood. But instead of the usual damp, musty, wet-dog smell of fresh-cut redwood, there arose a delicate, lemony, Pine-Sol scent. It didn't smell like redwood. But to the logger, it didn't quite smell like red cedar, either.

I met him at a pullout beside Highway 101. Together we walked switchbacked logging roads up from the shuttered mill. Regrown redwoods, Douglas firs, Sitka spruce, and western red cedar towered on either side. Finally we came to a landing. The retired logger shuffled through the underbrush until he found a collection of dark stumps, each as big across as a dinner table. These could be the strangely scented trees, he said. Back on that day in the 1980s, he'd hiked into the woods and marked the stumps with spray paint, intending to return and investigate further. But the mill was then a frenzy of activity, and there was no time. When the mill closed, the logger moved on to other things. We leaned over the stumps and sniffed. They didn't smell of anything but moss and decay. "If you had a chainsaw and cut into them, they'd smell," he said, but he was reluctant to be seen carrying a saw into what is now a protected state park.

The smell: Earlier, in his truck, he'd pulled out a Ziploc bag from the center console and passed it to me. It held a piece of wood, two inches thick, eight inches long, four inches wide, weathered gray on the sides. It was from one of the smelly redwoods, from a board he'd taken home and screwed to the wall of his garage. Over the decades the smell had faded from its surface. But when he cut away a piece, exposing a new orange-red face to the air, the smell reappeared. I opened the Ziploc bag and the truck was filled with a strong scent, broadly familiar—hops, weed, citrus, pine—yet exquisitely its own. I felt a strange sensation, unlike anything I could remember experiencing before, as though reality had rearranged itself around this little piece of wood. I could see how much was missing in my estimation of taxonomy—a sense of adventure, of discovery, of wonder, of awe, but also a hint of covetousness, of the pride that comes with knowing something almost no one else knows. It was intoxicating: There in my hand, maybe, maybe, was the most mysterious, dangerous, glorious kind of thing, a thing without a name.

~

Naming something raises it from obscurity into human terms, placing it on the map, giving it a lineage, turning it into something that can be returned to, discussed, studied, sought out, exterminated, conserved. Some of our oldest stories are about naming other species. In Genesis, God created the world and its creatures, every beast of the field and fowl of the air, and then He presented them to Adam to see what Adam would call them. "And whatsoever Adam called every living creature," Genesis records, "that was the name thereof." The account is vague on details, but it seems God did not explicitly give Adam the task of naming or even have need of these names Himself; instead, something in Adam compelled him to it.

"And Adam gave names to all cattell, and to the foule of the aire, and to every beast of the field," Genesis continues. But that is not quite true. Adam didn't finish the task, not even in his 930 years. Nor did his children, or his children's children. The task remained woefully incomplete by the mid-1700s, when Swedish botanist Carl Linnaeus styled himself a second Adam. Linnaeus provided a version of the descending scale of classification that taxonomists still use to arrange species on the tree of life, as well as the two-part naming system, also still in use.

By the time Linnaeus died, in 1778, he'd named some 12,000 species of animals and plants. His disciples scattered around the world, seeking species new to Western science, daubing them in Greek and Latin. Taxonomists have now named more than 1.5 million species and continue to describe about 17,000 new species every year. New species turn up in roadside ditches, behind supermarkets, in suburban New Jersey. More than one scientist I've talked with has discovered new species in their own backyard. Many of these species are minute or microscopic, but many others are of a size well matched with human perception. In recent years, scientists have described new whales, a new orangutan, even a new elephant. It is plausible, in other words, that something as conspicuous as a whole species of redwood tree has gone unnoticed and unnamed by Western science.

Life on Earth has the ungraspable, infinite quality of subatomic particles, of the ever-expanding universe, of the sea at dusk: The more progress taxonomists make, the bigger the task seems to grow. While some researchers estimate there are as few as two million species on Earth, and that Adam's task is therefore well underway, most estimates are much higher—that there are eight million species on Earth, or maybe two hundred million, or six billion. Using the last figure and the historical rate of discovery, to complete Adam's task would take us something like 350,000 years. When I started talking with taxonomists, I found that most of them seemed to view this situation less as an assurance of job security and more as an overwhelming problem. As climate change and habitat destruction threaten countless species with extinction—many of them, perhaps most, still unknown to us—taxonomists are increasingly engaged in a race against time. They will tell you how our taxonomic ignorance gets in the way of everything, dulls understanding, stymies accounting—how it is, when you think about it, among the biggest challenges facing humanity today.

This troublesome taxonomic ignorance is what I intended to write about, and I intended to write about it in an appropriately sober and serious tone. However, being easily distracted and of puerile taste and unfortunately interested in greed, pride, ineptitude, and other forms of human weakness, I quickly got caught in an eddy. My attention kept slipping away from taxonomy and toward the taxonomists themselves—and not the ones winning awards at conferences. Rather, I became fascinated by taxonomists like the one who named a dozen species of thunder and lightning, and the one who gave 248 different scientific names to what other taxonomists judged to be a single species of fly, and the one who decided to name the members of a genus of moths *Eucosoma bobana, E. cocana, E. dodana, E. fofana, E. hohana, E. kokana, E. lolana, E. momana, E. popana, E. rorana, E. sosana, E. totana, E. vovana, E. fandana, E. gandana, E. handana, E. kandana, E. mandana, E. nandana, E. randana, E. sandana, E. tandana, E. vandana, E. wandana, E. xandana, E. yandana,* and *E. zandana.*

I realized that the ratio of how important biological taxonomy is to how silly it is goes unmatched by any other branch of science. The reason is simple: Anyone can do it. Unlike in physics or physiology or phlebotomy, taxonomy remains open to everyone. This is a consequence of the rules that govern how taxonomists apply names to new creatures. Over the centuries, in order to keep track of millions of living things, taxonomists have bound themselves within a Byzantine system of rules, complete with its own kind of supreme court and legislative process. They view this system, known as "the Code," with a reverence that often borders on the religious, akin to how many Americans see the U.S. Constitution. These rules, though, have a curious side effect, which is that anybody, whatever their credentials or capabilities, can propose a new species. Then, if they have followed the rules, other taxonomists must at least consider their proposal. In other branches of science, the best scientists are free to ignore the contributions of the worst. In taxonomy, they must reckon with them.

There is a beauty to this arrangement. It means that, despite its history as a product and instrument of colonial-era Europe, the system of taxonomy that Linnaeus began is truly the prerogative of all humankind—it is our shared task, if we choose it. But collective endeavors always hover on the edge of chaos. Professional taxonomists, laboring together to impose order on an unruly world, have long complained about the cranks, kooks, and incompetents among them. In recent years, though, taxonomists have moved beyond pointing out mere incompetence or scientific wrongheadedness to accusing one another of a kind of crime: "taxonomic vandalism," in which a person describes and names numerous new creatures based on the thinnest of evidence, including evidence plucked straight from the papers of other scientists. Taxonomists around the world had fallen into debate over not only these alleged acts of taxonomic vandalism, but also the fierce reactions to them, both of which threatened to mire them all in confusion—to, as Genesis puts it in the story of Babel, "confound their language, that they may not understand one another's speech." It was in the context of this debate that I first heard of Raymond Hoser.

A taxonomist and blogger wrote a series of articles on what he called "the Raymond Hoser problem." There was an article about Hoser titled "A Few Bad Scientists Are Threatening to Topple Taxonomy." There were editorials about him in scientific journals and heated exchanges on taxonomic message boards. Everyone seemed to agree that Hoser, propelled by a rare combination of industry and shamelessness, was slapping names on creatures of dubious reality, and in such an overwhelming quantity that he threatened to turn huge swaths of the field of taxonomy into an unnavigable morass. Everyone also seemed to agree that it was an entertaining story. "As an outsider I find this thread fascinating," wrote a poster on one forum, "in a trainwreck kind of way."

I began lurking around Hoser's website, smuggled.com, which was filled with surprising pictures and bright flashing animations and colored and underlined and bolded and ALL-CAPs text, giving the impression of a traveling carnival. One evening I called one of the several numbers listed on the website. It was morning, Melbourne-time, and Hoser answered right away. "Helllll-o, Raymond speaking," he said. When I told him I was a reporter in the United States, interested in the topic of taxonomic vandalism, it was as if I had loosed the Aeolian knot. For more than an hour, without requesting my credentials or the terms of my employment and pausing only to take a call from a prospective client, Hoser offered an entirely different version of the events in question. He told me about Wolfgang Wüster and the other scientists who comprise the Wüster Gang, who he said were the real taxonomic vandals; about Wells and Wellington, whose controversial taxonomic treatises foreshadowed his own activities; about wildlife smugglers, rival snake catchers, corrupt members of the New South Wales and Victoria wildlife departments; about a meatpacking boss ground up and passed off as kangaroo meat. Most of all, he told me about snakes.

It can be hard to know what is true where snakes are concerned. People will tell all kinds of stories, and snakes themselves have something of a reputation as liars. "Do not believe those rigid threats of death," Milton's serpent said to Eve. "Ye shall not die: How should ye? by the fruit? it gives you life. . . . Reach then, and freely taste."

Snakes are mysterious, inscrutable, paradoxical, able to crawl without legs, to stare without blinking, to shed their skin and turn from old to young. As one writer observed, "They spend little time on love and none on parenting. They have no hobbies, make no art, and are not capable of returning the affection of those who claim to love them. If they philosophize, we have no way of knowing their thoughts or theories."

Snakes occupy a kind of border space in the cultural imagination, slithering back and forth between reality and myth. For instance, American dairy farmers once dreaded the milk snake, which would sneak into the barn at night and rob their cows, although distinguished herpetologist Clifford H. Pope doubted this. "There are no sucking muscles in its throat that would enable it to get the milk out," he noted, "for milking a cow is no simple task, as every milker knows." Colonial America was also home to the glass snake, which was prone to shattering; the whip snake, which would lash its victims; and, most fearsome of all, the hoop snake, which would take its tail in its mouth, form a wheel, and roll around at great speed. Its bite was deadly, capable of killing people, livestock, and thin-barked trees.

Of all the species of snakes that are not real, this hoop snake seems to be the most widespread. People in ancient China inscribed it on a clay jar some five thousand years ago. The ancient Egyptians carved a hoop snake on the walls of King Tut's tomb. Jesus, too, spoke of the hoop snake, as recorded in the Gnostic text *Pistis Sophia*. "The outer darkness is a huge dragon with its tail in its mouth," he told Mary Magdalene. "It is outside the world and surroundeth it completely." In the kingdom of Dahomey in West Africa, people believed that a hoop snake was a girdle around Earth's midsection, preventing it from bursting apart. People in Siberia also knew of the hoop snake. So did people in Italy, in India, in the Crimea, in Iceland, in Sumatra, in Benin, in the Peruvian Amazon. Scholars say that the ever-rolling hoop snake represented infinity and immortality, destruction and renewal, the narrow divide between chaos and order, darkness and light, beginning and end. Or perhaps it merely represents human nature, a universal joke about a squiggly line of a creature, a deadly beast whose fore so closely resembles its aft: Bite yourself!

"Everybody is interested in snakes, whether they like them or not," wrote the famous snake man Eric Worrell. More frequently, they do not. Ophidiophobia, the outsized, irrational fear of snakes, is one of the more common phobias. According to one study, it is present in 38 percent of women and 12 percent of men. Other primates are also afraid of snakes, with lab-reared monkeys somewhat less likely to be afraid of snakes than wild monkeys. Monkeys can acquire ophidiophobia simply by watching other monkeys be frightened by snakes, both in person and on TV. Researchers have discovered that people are far better at noticing animals and other humans in a scene than they are at discerning vegetation or inanimate objects, and that people are especially good at picking out snakes; ophidiophobes are even better at this task.

Ophidiophilia, a correspondingly outsized and often equally irrational passion for snakes, is less common. It is at the very least a contrarian position, and I have noticed that a love of snakes not infrequently coincides with certain philosophical tendencies, including a reflexive opposition, iconoclasm, even a sense of persecution. Perhaps the most fitting archetype is the Satan of Milton's *Paradise Lost*, who chose as his earthly vessel the serpent, the subtlest of all the beasts of the field. According to Milton's rendering, God cast Satan from Heaven not for sinfulness, per se, but for his indomitable spirit of defiance. Awaking in Hell, surrounded by his fellow fallen angels, Satan declared, "Here at least we shall be free. . . . Better to reign in Hell, than serve in Heav'n."

Not every snake man, of course, is this type. But the ones who are this type are almost exclusively men. In Australia, where snakes have long occupied an outsized place in the public imagination, there are and have always been snake women, but far fewer of them seem to have been arrested for public drunkenness or to have become embroiled in public controversy for selling fraudulent snake-bite antidotes or to have brandished their snakes against police officers. Relatively few snake women, furthermore, have been fatally bitten by their snakes. This same disparity is true within taxonomy. Although the proportion of female taxonomists is higher today than it has ever been, few if any women seem to have committed the kinds of heinous taxonomic acts that have frequently caused male taxonomists to go down to lasting infamy and derision. The single-minded and often delusional compulsion to name new species that sometimes infects taxonomists, meanwhile, seems to be predominantly a male disease.

In this intersection of taxonomists, agents of order and cooperation, and snake men, the embodiments of chaos and disharmony, I could not help but hear echoes of broader themes: of a growing distrust in institutions and experts; of the spread of dis- and misinformation; of my sense, more and more, that I am surrounded by people acting out their own private rebellions. It feels like a time not for building collective works but for tearing them down.

To me, this allegorical quality was alluring, offering rich opportunities for metaphor and symbolism and seeming to dangle the possibility of a work that was of not only educational but also literary or even artistic merit. Yet there were also times—like when I uncovered a centuries-old snake-bite-dueling conspiracy, or when a central character described to me what was apparently an abduction by beneficent aliens, or when I was an accessory to the theft of a large number of Chinese-language newspapers—when I wondered how the earnestness of my project's beginning had been subsumed by absurdity, and whether, in choosing to follow a story that, more than anything else, I found amusing, I was giving intellectual cover to what was at its heart mere voyeurism.

Indeed, as I began retracing the story, a number of people suggested I was better off not writing about Raymond Hoser and the other snake men. Even though nearly all of it had already appeared in news stories and court files and other primary documents, they said that in compiling and synthesizing those scattered reports into a cohesive narrative, I was inviting trouble, rewarding attention-seeking behavior, resurrecting a name that had deservedly been struck from official memory—even that I was contributing to the problems facing the science of taxonomy itself. Perhaps that is true, but caught up as I was in a story of rebellion, I chose not to listen.

So it was that, nearly two years after I first spoke with Hoser on the phone, I wound up in Australia, perched by the corner of his desk, watching as he stood at the sink swabbing snake shit with one hand and with his other hand juggling a succession of the world's deadliest snakes—tiger snakes, black snakes, inland taipans—which he paid almost no attention. He was still busy telling me his side of the story, which was relatively straightforward: Wolfgang Wüster and his colleagues were the ones who had committed taxonomic vandalism, he said, and they had done it on an astonishing scale. Their motivation, he said, was simple jealousy. They, the academy- and museum-based professional taxonomists, had been outdone by him, the Snakeman, for whom taxonomy was only an unpaid side gig, a passion project. By his own count, Hoser had named more than a thousand new species and subspecies and another thousand genera, tribes, families, and so on, making him the most prolific taxonomist in living memory.

Now, he said, his enemies were at his gate, trying to ruin him, not only squashing his taxonomy but tampering with his snake-catching business, sullying his reputation, sending rival snake catchers and performers against him, messing with his standing on internet search engines. Worst of all, in their relentless campaign against Hoser, they had disregarded the centuries-old system of rules that direct how species are named, therefore threatening the stability of biological taxonomy in general, and therefore conservation efforts around the world, and therefore the survival of countless species, and therefore, you might argue, the future trajectory of life on Earth itself.

He paused. Clutching the pair of eastern browns, he crossed the room, pulled *Snakes of Western Australia* down from the shelf, and opened it in front of me. I recognized the snake he was pointing to, because, while some of the story goes back many decades or even centuries, it could also be said to have begun with the snake in the book, commonly known as the Pilbara death adder. The snake was red-orange, with a triangular head, latitudinal zigzag stripes, and the thick body and sharply tapering tail of a fresh tube of toothpaste. It was the first species he'd ever named. There in the book, next to the snake's common name, was the two-part scientific name he'd given it, which, per the usual taxonomic practice, was followed by his own name and the year he'd named it: *Acanthophis wellsi* Hoser, 1998. The link between the snake's name and his own was unassailable, eternal, indelible, he told me as the brown snakes wriggled around my ankles. In the name lay a kind of immortality. Whatever insults and injustices his enemies might visit upon him, he said, "They'll never take that from me."

AUSTRALIA TERRA COGNITA

I tell you zat I had no thought
To say zat vot I did'nt ought;
It vos no part of mein own wishes,
To miscal men while naming vishes.

On April 19, 1770, the HMS *Endeavour* sailed into sight of a small protrusion on Australia's southeastern coast. "I have named it Point Hicks," Captain James Cook wrote in his log, "because Lieutenant Hicks was the first who discover'd this Land."

This was not really true. Dutch sailors had visited Australia in 1606, making landfall at what is now the Cape York Peninsula, on Queensland's far northern coast. Over the following century and a half, more Dutch and French expeditions traveled along the Australian coast. The area was also well known to local inhabitants. Such technicalities did not concern Cook. Like many European explorers of the day, he was possessed of an imagination that allowed him to see a place that was full of people as functionally empty and a place where most everything already had a name as in need of new and better ones. As the ship followed the shore, he set himself to the task. He called a peak Mount Dromedary for its camel-like figure. He named another point of land Point Upright, "on account of its perpendicular

Cliffs." Yet another point he called Red Point, because "some part of the Land about it appeared of that Colour." He spotted "a remarkable peak'd hill laying inland, the Top of which looked like a Pigeon house." He called it Pigeon House.

Expedition naturalist Joseph Banks was also busy naming. In his journal, he noted the creatures he'd snagged with his dip net: *Actinia natans, Dagysa corunata, Mimus volutator, Dagysa gemma, Medusa pelagica, Helix janthina, Doris complanate, Beroe biloba, Cancer erythroptamus, Holothuria obtusata,* and *Phyllodoce velella.* Most of these names, attached to creatures known more commonly as sea anemone, marine worms, jellyfish, crabs, snails, and sea cucumbers, seem to appear only in Banks's journal, suggesting he bestowed them himself, likely intending to follow up later with formal descriptions. There were so many new species that he didn't have time to do anything but give them names.

For most of history, biological taxonomy did not require its own science. It was an integral part of the human experience, something that children learned from an early age. Its broad outlines were obvious. Most people will agree, for example, that there are things that are alive and things that are not alive, such as rocks. Among living things, the most obvious division is between animals and plants. Within those categories, there are oft-repeated architectures, like birds, trees, and fish. There are kinds of birds and trees and fish: waterfowl and conifers and flatfish, for example. There are kinds of these kinds, too, like swans and firs and flounder. Anthropologists note that these rough linguistic and conceptual categories are consistent across the world's cultures, notwithstanding occasional differences of detail. (As Carol Kaesuk Yoon writes in *Naming Nature,* for instance, the Rofaifo people of the New Guinea highlands include the cassowary in their "large mammal" category, despite the cassowary possessing wings, a beak, and feathers.) This consistency of concept matches consistency of reality, writes anthropologist Brent Berlin: Whereas people are free to construct different rituals, religions, and artistic traditions, "human beings everywhere are constrained in essentially the same ways—by nature's basic plan—in their conceptual recognition of the biological diversity of their natural environment." People, Berlin continues, "do not construct order, they discern it."

The roots of this taxonomic ability must lie deep in our evolutionary past—surely every trilobite could distinguish between predator and prey. Prairie dogs use different chirps for different types of predators. Vervet monkeys scream when they see a snake, cough when they see an eagle, and bark when they see a leopard. The taxonomies used by traditional societies are equally practical. Creatures that are distinct and useful get names, Berlin writes, while less obvious or consequential creatures go nameless. Traditional societies, anthropologists have found, tend to use around six hundred taxonomic names, mostly at the level of genus.

By the 1700s, though, European naturalists were devoted to a grander vision: naming every living thing and arranging it in relation to every other living thing. This task, unfortunately, was growing more ponderous by the day, as explorers sailed the world, dragging back thousands, then tens of thousands, of life-forms unlike any found in Europe. As naturalists struggled to arrange these creatures in relation to one another, they also struggled with how to name them so that everyone would know which creature was being discussed. Some naturalists tried to nestle definitions into names. Common yarrow, by this method, became *Achillea follies duplicatopinnatis glares, laciness linearibus acute laciniatus.* The common European honeybee became *Apis pubescent, torhace subgriseo, abdomen Fusco, pedibus posticis glares utrinque margine ciliatis.* Taxonomy, once practical, so much a part of life as to go unnoticed, had become a logistical nightmare.

Linnaeus cut through the jumble. In his 1735 paper *Systema Naturae,* he provided a descending taxonomic scale of relation. It mirrored earlier folk taxonomies, separating the kingdoms of animals and plants (and minerals), then classes, orders, genera, and finally species. Then, in 1753, in *Species Plantarum,* he offered a method of naming, with each species receiving a two-part name composed of genus and species. These names could be descriptive but didn't need to be. By Linnaeus's system, the two-part species name was also joined to the name of its namer. So the Swede affixed his name to ours: Rendered in full, our species is *Homo sapiens* Linnaeus, 1758. This last detail was perhaps more important than it first appeared, setting up many of the conflicts to follow: It conferred a measure of glory.

Later, people gave Linnaeus the honorific *Deus creavit, Linnaeus disposuit*—Latin for "God created, Linnaeus organized." For his part, Linnaeus saw his efforts as a tribute to God. But a person might also argue to the contrary, that Linnaeus's project stood in defiance of God, who long before had moved to prevent such a thing. The story is that, after God cast Adam and Eve from the Garden, after He set the mark upon Cain, and after the waters rose and receded, the people of the world gathered together and said to one another, "Let us build a city and a tower, whose top may reach unto heaven; and let us make a name, lest we be scattered abroad upon the face of the whole earth." But such ambition was an affront to God, so He scattered the people and forked their tongues in all directions so they could no longer understand one another. They went forth and named Earth's creatures in their thousands of languages, and often they named the same life-forms more than once, and it was not at all clear which was which.

But then came Linnaeus with a plan to build a great tower that had ziggurat steps made of every living thing, and he drew many people to his vision, and now when they scattered across the Earth, they took with them a new language that they all shared.

On April 28, 1770, the *Endeavour* sailed into what Cook called Botany Bay, in what is now known as Sydney, Australia. There, its sailors encountered a group of about ten people. These people did not have written language, so details of the encounter come primarily from the journals of the sailors, including that of Joseph Banks, who called the people "Indians," using the term even more recklessly than Christopher Columbus had three centuries before. When some of the sailors rowed ashore, two men from the group met them, shouting and brandishing spears. "They remained resolute so a musquet was fird over them," Banks wrote. This did not scare the men off, so the sailors shot one of them in the legs with bird shot. "He minded it very little so another was immediately fired at him," he continued. "On this he ran up to the house about 100 yards distant and soon returned with a shield."

In their own language, the people were Dharawal, just one of hundreds of groups of people scattered across the continent. Many of these groups shared a similar story of how they had come to the place: They had always been in the land, existing in a kind of eternal state without beginning, and then one day, as the poet Oodgeroo Noonuccal recounted, "the Rainbow Serpent awoke from her slumber and pushed her way through the earth's crust, moving the stones that lay in her way." The Rainbow Serpent traveled all over the world, carving valleys and casting up mountains. Then she summoned frogs from their beds beneath the soil, and when she tickled their bellies, water ran from them and filled the Rainbow Serpent's tracks. Plants grew, and the rest of the animals awoke.

Archaeological, linguistic, and genetic evidence suggests that people arrived on the land as early as sixty-five thousand years ago. These people would have recognized some of the animals and plants they encountered, but many would have been wholly new. Some 80 percent of species in Australia occur only there. There were giant wombats and marsupial lions, enormous kangaroos and twenty-foot monitor lizards and huge flightless birds in the chicken family. Traveling through a land that was truly empty of names, the people might have behaved much as Cook and Banks did millennia later, giving what historian George R. Stewart called "bestowed names"—that is, names joined to a place or thing purposefully and at a specific point in time.

Over time, the toolmarks of bestowal weathered away. Why was that place called that? Why was that creature called that? No one could remember. The land and its living inhabitants came to be known by what Stewart calls "evolved names," whose origins are no longer clear. The people multiplied, spread out. Different groups of people developed different accents, different grammars, and different words. By the time Europeans arrived in Australia, Aboriginal people spoke as many as three hundred different languages. Here, the stories of the Dreamtime and of Babel and of Western science come into alignment: There was a vast continent, full of people and culture and language and knowledge, and it was teeming with names.

After Captain Cook's violent encounter with the Dharawal, which unfortunately set the tone for subsequent relations between European colonists and Aboriginal people, he claimed New South Wales for the king. The *Endeavour* continued along Australia's east coast, lodged itself on the Great Barrier Reef, and eventually proceeded on to England. Banks and his fellow expedition naturalist Daniel Solander had collected some 1,400 new species of plants and animals. They instantly became celebrities, going on to illustrious careers. Cook, meanwhile, set out on a second, then a third voyage, eventually dying in the Hawaiian archipelago, which he called the Sandwich Isles, victim to what some scholars contend was a taxonomic mistake—the Hawaiians, by this version, mistook Cook for the god Lono and killed him ritualistically. More recent analysis suggests that the Hawaiians killed him in the usual sense, motivated by what was in fact intense dislike.

In 1787, Great Britain sent its First Fleet to settle the area around Botany Bay, where Cook and Banks had encountered the Dharawal. The Second and Third Fleets followed soon thereafter. The colonists multiplied, spread out, bestowing more names on their surroundings, often using what felt like the closest equivalents from the home islands. They also adopted words from the various languages of the Aboriginal people. The resulting linguistic amalgamation is well illustrated by the common names for the seven-hundred-odd members of the genus *Eucalyptus*, which dominate forests across Australia. During the *Endeavour*'s approach, Banks noted trees that drooped "like those of the weeping willow." When he visited them up close, he found that they oozed a sticky, gummy sap, so he called them gum trees. Today, other common descriptors include stringybark, paperbark, ironbark, pricklybark, kindlingbark, candlebark, scentbark, half-bark, mahogany, ash, redwood, peppermint, messmate, box, gimlet, tingle, yellowjacket, woollybutt, blackbutt, marlock, moort, mallet, mallee, sallee, wandoo, and yate. The eucalyptus that the jolly swagman sat under while boiling his billy beside the billabong was, of course, a coolibah.

Scientific exploration of the continent proceeded more slowly. In his history *Animal Nation*, Adrian Franklin quotes from the records of a Royal Society of New South Wales meeting held in 1821: "Upwards of thirty years have now elapsed, since the colony of New South Wales was established in one of the most interesting parts of the world—interesting as well from the novel and endless variety of its animal and vegetable production, as from the wide and extending range for observation and experiment. . . . Yet little has been done to awaken a spirit of research or excite a thirst for information amongst the Colonists." For a long time, writes ecologist Paul Humphries, naturalists stuck close to the coast, deterred from venturing inland by the lack of water and by the presence of potentially hostile Aboriginal people. "It was much more convenient to throw a net over the side of a sailing ship or collect along the seashore than it was to venture inland into inhospitable territory," Humphries writes. One of the first systematic biological surveys of Australia's interior came in 1856 and was led by the talented and much-maligned Johann Wilhelm Theodor Ludwig von Blandowski.

~

Born in 1822 in what is now Germany, Blandowski was the thirteenth and youngest child of an army officer and his wife. He was thirteen when his father died, which some biographers have suggested was the root of his later difficulties. Whatever the cause, his contemporaries agreed on his impatience, irritability, and tendency toward daydreaming and grandiosity, and also on his impossibly thick accent. In various photographs and portraits, Blandowski is handsome with long dark hair, unkempt sideburns, and a beard. He usually looks annoyed.

In his mid-youth, he dropped out of school, finding work in the region's coal mines. He proved himself exceptionally capable, bound for success, until one day he was spotted in association with protesters against the government and was forced to resign. "For some time, I have been considered by my superiors and also the inhabitants of Königschütte as a *rebel*, and I seem to upset their nightly peace," he declared in his letter of resignation. With bleak prospects in his home country, Blandowski traveled to Australia, to the settlement of Melbourne in the territory of Victoria, where he fell in with a group of German-speaking naturalists. In the years that followed, Blandowski drifted between jobs, working as a field collector for naturalists and as a geological surveyor. He tried his luck in the Victorian goldfields before returning to Melbourne, where he launched several short scientific expeditions. "One constant theme is that the horses are lost," recalled one man who accompanied Blandowski on an early expedition. "Suppose him standing over the fire deep down in the consideration of some philosophical chimera, or perhaps of nothing, when suddenly he rouses up stares across the gully, turns round as if on a pivot and shouts out 'Gentlemans the horses am walk off.'"

In 1854, the government of Victoria made Blandowski the curator of the new Victoria Museum of Natural History, despite certain hints that he lacked the organizational and interpersonal skills requisite to the job. Almost immediately, funding ran out and Blandowski was relegated to the status of collector. After a year of inactivity, he convinced his superiors to put him in charge of a cross-continent expedition to explore the biological riches of Victoria's interior, following the Murray–Darling river system north from Melbourne. The expedition set out on December 6, 1856. Along with Blandowski, it was composed of twenty-six-year-old natural-history illustrator Gerard Krefft, taxidermist James Mason, and Messrs. Weitenau and Dorgerloh, credentials unknown. They had "several drays and various recalcitrant horses and bullocks," one biographer wrote, "which were the bane of the expedition." The animals helped them carry "an enormous amount" of bedding, clothes, and food, a whole library's worth of books, preservative for specimens, guns, and more than six hundred pounds of photographic equipment.

For weeks, Blandowski's expedition wended its way northwest. The tedious work of guiding a train of horses and bullocks did not seem to suit Blandowski. He often rode ahead, sleeping at inns while the rest of the expedition straggled along behind and camped beside their animals. Finally, four months after leaving Melbourne, the group arrived at a small hill overlooking the Murray River, a place the local Nyeri-Nyeri people called Mondellimin. There they set camp.

Blandowski arranged with the Nyeri-Nyeri to trade animals they brought him for flour, tea, sugar, and other goods. In a land where nearly everything was new to them, the Europeans were quickly inundated with specimens. "An empty gin case constituted my table and a keg half full of decaying reptiles . . . served as a seat," Gerard Krefft, the illustrator, wrote in his account of the trip. "The stench of my seat and the great many carcasses of birds and animals lying about not only attracted but bred any number of flys and other insects. . . . I could use only one hand when I was drawing as the other was kept sufficiently busy to prevent the flys from blowing my eyes."

Blandowski had little patience for this type of work. "He seemed to see himself more as a heroic explorer than as the leader of a collecting expedition," one historian wrote. Twice he set out alone on horseback into the bush, once for three weeks, the second time for a month. During the second trip, while standing at the summit of Mount Jamieson, some four hundred miles north of Melbourne and nearly two hundred miles from the expedition's camp at Mondellimin, Blandowski took in a grand panorama. "The whole horizon was closed in with high blue mountains and picturesque hills," he wrote, "and my feelings then can only be understood by one, who himself has been on the verge of civilization . . . and gazed into the unknown wilds expanding before him."

In Patrick White's 1956 book *Voss*, the title character is a German leader of a cross-Australia expedition who similarly sets off alone into the wilderness. The character was modeled after Ludwig Leichhardt, another German who made an exploratory expedition across New South Wales in 1848, and who was known to have also wandered away from the main body of his expedition. Voss the character and Leichhardt the man both died during their expeditions, and their desert wanderings made them seem mysterious, even messianic. Conversely, despite Blandowski's surviving his own expedition, his sojourns made him seem neither mysterious nor Christlike. Gerard Krefft was especially unimpressed.

Upon returning to Mondellimin, Blandowski determined that Krefft was carrying out his orders adequately, so he left once more, this time catching a steamboat downriver to Adelaide, then on to Melbourne. At first, his sponsors judged the expedition a success. He "appears to have been received as a minor celebrity on return to Melbourne," one biographer wrote. "Within two weeks, Blandowski's backers had arranged a celebratory dinner in his honour." Another biographer wrote that this "was undoubtedly the highlight of Blandowski's career."

But soon after, in an incident that both spoiled Blandowski's spell of good fortune and illustrated the potential of Linnaean taxonomy to serve as a means of settling scores, he submitted to the Victorian Philosophical Institute a paper with descriptions of nineteen species of fish that he claimed to have discovered during the expedition. These included what he called *Cernua eadesii*, named for institute cofounder Richard Eades, which he described as a "fish easily recognized by its low forehead, big belly and sharp spine," and *Brosmius bleasdalii*, named for Rev. Dr. John Bleasdale, another institute member, which he described as a "slimy slippery fish. It lives in the mud and is of a violet bluish colour on its belly."

The institute erupted into argument. Some members sided with Eades and Bleasdale, the fishes' namesakes, who felt that Blandowski's descriptions were intentionally barbed. Others argued it had been an honest mistake, attributable to Blandowski's poor English and brusque manner. Gerard Krefft, seeking to distance himself from his impolitic boss, noted that Blandowski had not been present when the two offending fish were collected, and therefore could not have known their habits. *Melbourne Punch* gleefully followed the story, publishing a fifty-six-line poem that included the stanzas:

Mein himmel—all vot I wish,
You look yourself upon ze vish;
You see ze spines—ze little pins,
Vot sticks out here upon his fins.

This oderr vish has got ze same,
But he has got anoder name
Because he differs in his belly,
Vich is all vlat, not round and swelly.

Blandowski's reputation was tattered, his ambitions thwarted. It was the kind of frustration that has often fired rebellious souls to even more dramatic acts. But Blandowski chose to withdraw. He returned to Germany, where he moved back in with his sister and mother, taking work as a photographer, a profession at which he found little success. For years, he had planned to publish a book entitled *Australia Terra Cognita*—Australia, the known land—which he hoped would find company with the grand travelogues of the day, like Humboldt's and Darwin's descriptions of their journeys through South America or Linnaeus's account of his trip through northern Scandinavia. He enlisted an art student to create illustrations from the sketches he and Gerard Krefft had made during their expedition. One of the finest of these illustrations portrays Blandowski on a beach, wearing a wide-brimmed hat and raising a whip, engaged in deadly combat with a seal. In September 1873, he was admitted to the Silesian Provincial Mental Asylum and died that December, age fifty-six.

After languishing for a century and a half in ignominy, Blandowski has recently enjoyed something of a renaissance. In 2007, he was the subject of a symposium that resulted in a two-hundred-page book in which a dozen scholars examined him through their scholarly lenses. Blandowski, they argue, was a notable and unusual historical figure for several reasons. He was the first in the region to carry out a scientific expedition of its type, for one. The notes he and Gerard Krefft took on the various animals they encountered provide a useful record from an era of rapid change; the region would soon be overrun with white settlers, who unleashed ravenous hordes of cattle, sheep, and rabbits. Similar wildlife surveys wouldn't be resumed in Victoria for more than a century. By then, one historian writes, many of the animals Blandowski and Krefft encountered were gone from the region.

The modern scholars also applaud Blandowski for his relatively enlightened approach to Aboriginal people. Enlisting the help of the Nyeri-Nyeri in collecting species was "a masterstroke," as one of those scholars put it. This seems a bit like congratulating someone for stopping to ask for directions, but it's true that, among his contemporaries, he was unusual in seeing Aboriginal people as potential partners. Other naturalists and taxonomists of the era were more likely to view them as scientific subjects. *Endeavour* naturalist Joseph Banks, for instance, kept a collection of human skulls, including some belonging to Aboriginal people, and he enthusiastically helped other naturalists acquire similar collections. White colonists in Australia routinely murdered, lynched, and raped Aboriginal people, and only rarely did they face trial. By these standards, Blandowski was indeed progressive. In his published report on the Murray–Darling expedition, he thanked his Aboriginal collaborators, "to whom I am indebted for all the information and discoveries I have made." One historian suggests that Blandowski, ever the contrarian, may have viewed the Aboriginal people he encountered, marginalized on the lands they'd occupied since time immemorial, with a sense of kinship, as fellow outcasts.

Despite Blandowski's generally favorable reappraisal, his legacy as a taxonomist could not be redeemed. Other naturalists immediately derided the specimens he and Krefft collected during their expedition, noting that they were of poor quality, were badly labeled, and contained an enormous number of duplicates. "Sadly they disappoint the naturalist," wrote one former colleague, describing the eighteen thousand shells that Blandowski had collected, "for the whole number only includes about half a dozen species, and the contribution might be roughly described as a bushel of mussels and another of periwinkles." Later, an independent taxonomist and natural-history dealer commissioned by the museum estimated that all of the samples collected during the trip were worth a total of £313, a small fraction of what the expedition cost. Of the nineteen species of fish Blandowski named, ten were found to be duplicates, several were already named, and one was discarded. By the time of his death, Blandowski's mark as a taxonomist was almost entirely erased.

Gerard Krefft, meanwhile, went on to a distinguished but brief career. In 1862, the governor of New South Wales appointed him assistant curator of the Australian Museum, and he soon rose to curator. During his time in the position, he named numerous species and wrote various taxonomic works, including his 1869 book *The Snakes of Australia*. In the preceding quarter century, scientists had described some fifty Australian snakes, bringing the total to nearly seventy. Krefft himself had named fifteen new species. The remainder were named by just a handful of men, most of whom never visited Australia. Almost all of the names they gave the snakes, both scientific and common, were conspicuously of European origin.

In the early 1800s, for example, naturalist François Marie Daudin described a snake collected during a French expedition. "Body and tail somewhat elongated, cylindrical," Daudin wrote. "The rounded, obtuse head, slightly domed over the eyes, covered in front with polygonal plates, smooth and few in number. . . . Jaw teeth sharp, curved. Poison fangs?" He noted a spike on the tip of its tail—"fine, small, compressed, very sharp, slightly curved towards the sky"— which inspired the name he gave the new genus: *Acanthophis*, made up of *acanth*, from the Latin for "spined," and *ophis*, from the Greek for "snake." In the same document, Daudin also gave the animal a species name, *cerastinus*. But the same species had been described a year earlier by George Shaw and E. R. Nodder, who wrote that "this beautiful, and hitherto undescribed serpent, is a native of Australasia, and is considered a highly poisonous species." They called it *Boa antarcticus*, or "Southern boa," despite it being clearly not a boa. In accordance with standard practice, taxonomists later combined Daudin's genus with Shaw and Nodder's species name. The animal appeared in Krefft's *The Snakes of Australia* as *Acanthophis antarcticus*.

Death adder, the common name that Krefft recorded, was of similarly convoluted origin. The English word "adder" comes from the Old English *nædre*, and refers specifically to the European viper, which Linnaeus named *Vipera berus* in 1758. The word was often used more generally, though, as a synonym for "snake" or "serpent," as in Psalm 58, which compares the wicked person to "the deaf adder that stoppeth her ear; which will not hearken to the voice of the charmers." This is how the Australian snake was described in the first clear reference to it, in an 1827 article in *The Australian*. It appears that colonists had applied the name "adder" to the Australian snake in the general sense, the same way that they tended to call the conifers they encountered "pines," and the hardwoods "oaks." "Adder" was in this case very general. The European viper, which has a slender body, a teardrop-shaped head, and a longitudinal, zipper-like dark stripe down its back, bears little resemblance to the Australian snake, with its stout body and tapered tail, spade-shaped head, and latitudinal stripes. Genetic studies suggest the two snakes went their separate evolutionary ways some 50 million years ago, at a time when whales were still galloping around on cloven hooves.

The snake continued to go by "deaf adder" until 1845, when it was described by *The South Australian* as the "death adder," an apparent malapropian shift. For the next century, the bestowed name and the evolved name competed for dominance, with "death adder" slowly winning out. This linguistic tug-of-war is not mentioned by Gerard Krefft, who a few years after publishing *The Snakes of Australia* came to disgrace when museum trustees accused him of stealing gold from the museum and of using its printing press to publish pornography.

So the snake, neither adder nor boa, acquired both a bestowed name and an evolved name, and the older names people had given it in many languages were mostly forgotten. It became one more brick in Linnaeus's great tower, which promised to bring order out of the chaos of life on Earth. But as the tower grew higher and higher, its builders noticed a problem, embedded in its very foundation, that threatened to topple it.

CHAPTER 3

THE MIHI-ITCH

In those days there was no king in Israel: every man did that
which was right in his own eyes.

—JUDGES 21:25

The trouble had started almost immediately: As European naturalists scattered around the world, applying Linnaean binominals to the creatures they discovered, they often named the same creature more than once. Sometimes this was an honest mistake, a question of logistics—the distances were great and the ships slow, and while the scientific names were given according to a common system, they appeared in papers written variously in English, French, German, and other tongues. But not all of the double-naming was unintentional. As Hugh Strickland wrote in 1835, it was apparent that many naturalists were renaming species on purpose. At best it was vanity; at worst, outright theft. Anarchy loomed, Strickland warned— and perhaps even the end of Linnaean taxonomy altogether.

Strickland, born in 1811 to a wealthy family on England's eastern coast, loved rules and order and loathed deceit and disharmony. His father-in-law and biographer later reported that Strickland possessed "an instinctive distaste for all exaggeration . . . in relating any event or circumstance he made a point of being strictly accurate." Indeed, in his memoirs Strickland exhibited a complete lack of the storyteller's guile.

He was at home, instead, in the fact-bound world of science. When he was sixteen, he published his first scientific paper, about potential mechanical improvements to the weathercock, and he followed this paper with a prodigious string of others, on such topics as Egyptian hieroglyphics, methods of storing seeds, the orbital nature of shooting stars and other celestial bodies, an early photocopying technique, the placement of accent marks in prose translations, and the luminosity of glowworm eggs. Most of all, he wrote about his two passions: geology, which led him eventually to his shocking and untimely death, and birds.

The scientific bird literature, to Strickland's ongoing dismay, was chaotic, disordered, rotten with so-called synonyms, where naturalists had given more than one name to what was, in fact, the same species of bird. He estimated that there were at least three times as many scientific names as there were actual species. Naturalists renamed species and renamed them again, all in accordance with the systems that they each seemed to have developed for their own use. As Strickland and a group of colleagues later put it, "One author lays down as a rule, that no [species] names should be derived from geographic sources, and unhesitatingly proceeds to insert words of his own in all such cases. Another declares war against names of exotic origin, foreign to the Greek and Latin; a third excommunicates all words which exceed a certain number of syllables; a fourth cancels all names which are complimentary of individuals; and so on."

There are valid reasons to rename species. To describe a species is to offer a hypothesis, proposing that the species is, in fact, a meaningful entity, a group of individuals like one another and distinct from other groups of individuals. This hypothesis might be supported by various lines of evidence—by the purported species' physical form, or its genetics, or its habits, or its favored type of habitat, or its location on Earth. Based on the strength of this evidence, other taxonomists might accept or reject the hypothesis. In the process of offering new hypotheses or reevaluating old hypotheses, taxonomists will often rename species.

But as Strickland complained in a letter published in 1835, when he was twenty-four, his peers were renaming species not for scientific reasons but for nomenclatural ones. Nomenclature is the system of naming that Linnaeus proposed and that still governs how taxonomists apply names to the creatures they describe. In essence, nomenclature is a filing system, used to keep track of what has been named, when, and by whom. In theory, the names that taxonomists give to creatures are mere headings in this filing system, a way to ensure that everyone is talking about the same type of creature. The meaning of a name is therefore almost an afterthought, to be viewed with an agnostic detachment. But naming something is an act of creation, a moment when the namer forever affixes their name to that of their discovery. It is an opportunity to pay favors, to settle accounts, to make a joke, to stoke the ego. It might even bring the taxonomist attention from the world beyond the narrow halls of their science. Perhaps it was inevitable, then, that taxonomists seemed unable to resist naming—or renaming—creatures for reasons related not to the science but to the meaning of the names themselves.

In his letter, Strickland pointed to what he saw as an especially egregious example: The year before, in the *Magazine of Natural History*, a person identified only as SDW had proposed to rename the common bullfinch, a small songbird with banded wings, a black cap, and, in adult males, a red-orange breast. In 1758, Linnaeus had given the bird the binominal *Loxia pyrrhula*, derived from the Greek for "flame-colored bird," placing it in a genus with the crossbills. Two years later, another naturalist argued convincingly that the species belonged in its own genus, calling it *Pyrrhula pyrrhula*, as it had remained ever since. SDW now offered a completely different scientific name, *Densirostra enucleator*, which Latin-speaking nineteenth-century naturalists would have understood to mean, roughly, "a bird that uses its heavy beak to open seeds." SDW wrote that this new name was "more definite and expressive," and suggested that a better common name for the bird was "coalhood thickbill."

Strickland was outraged. The whole point of Linnaeus's binominal system of nomenclature was to free taxonomists from the need to assign names that precisely described the creatures in question, he wrote in his letter, also published in the *Magazine of Natural History*. "As we do not object to *William Whitehead's* name because his hair may happen to be *red*, so, if the *meaning* of a specific name be downrightly inapplicable to the object, this need not prevent its *sound* being adopted as the conventional sign of the species," he wrote. The important thing, that is, was that everyone agree that a given name referred to a given creature.

While turning the common bullfinch into the coalhood thickbill might have seemed like an insignificant act, Strickland wrote, it was prioritizing the nomenclatural filing system over the science of taxonomy. It was the abandonment of data, evidence, and objectivity in favor of personal whims and ideals. It was glory-seeking. It was lawlessness. It was the first step on the road back to chaos. "If this practice be once given way to," Strickland wrote, "there will soon be an end of all nomenclature, and, through it, of all science." He continued, quoting Linnaeus: "For true it is, that, 'Nomina si pereunt, perit et cognitio rerum'—If names perish, the knowledge of things perishes with them."

~

What happened next is what usually happens when a little-known twenty-four-year-old writes an editorial proposing that his older and more famous professional colleagues adopt sweeping changes that would restrict their freedom: They ignored him.

"The lovers of confusion have been hard at work," Strickland wrote in a second letter, published in 1837, complaining that those few colleagues who deigned to respond had done so "in not the most courteous tones." Undeterred, he offered "a few general rules," twenty-two nomenclatural commandments that he hoped might guide Linnaeus's disciples in the process of naming the world's creatures. Some of these rules addressed punctuation and minor questions of etiquette, but the important ones concerned what became known as the "principle of priority." The person who first described a species, Strickland wrote, should get to name it, and once named, a species should be renamed only for taxonomic—that is, scientific—reasons, not nomenclatural ones.

Once again, Strickland's peers mostly ignored him. Strickland was not swayed—after all, Linnaeus had also been a little-known naturalist in his twenties when he'd published the first edition of *Systema Naturae*. Strickland began a more concerted lobbying campaign, writing letters and editorials and arguing his case with everyone he met. At an 1841 meeting of the Zoological Section of the British Association for the Advancement of Science, then the leading scientific organization in the world's leading colonial power, Strickland once again proposed nomenclatural rules to his peers. Association members declined to take them up at that meeting, but later that year, Strickland gained permission to assemble a committee to discuss and offer recommendations on the topic. This committee presented its report at the association's next meeting, in June 1842. After Strickland read the report aloud, his father-in-law later recounted, "it encountered an opposition that was scarcely expected, couched in a spirit of prejudice, and almost jealous animosity." Some naturalists denounced the rules as unnecessary or misguided; others, in a scientific era dominated by wealthy "gentlemen," deemed them insulting. Few seemed eager to voluntarily adopt restrictions on how they named species.

But while the report was "well thrashed," Strickland's father-in-law wrote, critics of Strickland's rules could not muster a unified opposition, and his key point proved irrefutable: Linnaean taxonomy, which was an effort to impose order on the world, had itself grown disordered. Strickland convinced the British Association for the Advancement of Science leadership to print the rules in its report on the 1842 meeting. Momentum was now in Strickland's favor, and in 1846, more than a decade after Strickland began his effort, the British Association formally adopted the nomenclatural rules, joining several scientific societies that had already done so.

It looked like success, at first—Strickland, through his patience and determination, had rallied eminent naturalists and the world's leading scientific society to his vision of consistency and rectitude.

But the rules, which could only persuade, not compel, failed to find traction, especially abroad. Even his converts lacked fervor, often ignoring the rules when they proved inconvenient. In an 1849 letter to Strickland, Charles Darwin admitted that, while he found it "quite a comfort to have something to rest on in the turbulent ocean of nomenclature . . . I find it very difficult to obey always."

In the face of widespread indifference, Strickland turned back to his twin passions. Over the next decade, he wrote only rarely about nomenclature, instead publishing numerous articles and treatises on geology and birds, including one on the Mauritanian dodo, which, as he and his coauthor wrote, was the first clear instance of humans driving a species to extinction. Many other species of animals and plants "are now undergoing this inevitable process of destruction before the ever-advancing tide of human population," they observed. "The Zoologist or Botanist of future ages will have a much narrower field for his researches than that which we enjoy at present. It is, therefore, the duty of the naturalist to preserve . . . the knowledge of these extinct or expiring organisms, when he is unable to preserve their lives."

One of Strickland's favorite pastimes was visiting railroad cuts, where layers of rock lay exposed. His father-in-law later wrote that Strickland usually visited these cuts as part of a group and usually checked with local stationmasters for the train schedule. But on September 14, 1853, he was alone, and no stationmaster was on duty at the town of Retford. He walked along the rails to Clarborough Tunnel, bored a few years before. Witnesses say that Strickland was so focused on the rocks in front of him that as he stepped back to avoid a coal train steaming toward him on one track, he seemed not to notice a passenger train bearing down on the other.

~

By the time of Strickland's death at age forty-two, his system of rules had foundered. More than a decade later, his father-in-law summed up the situation, writing that "nomenclature has not improved. Whether it is from the rules and recommendations not being sufficiently well known, or from an idea that no one has any right to interfere with or make rules for others, many gentlemen appear to cast them away, and do not recognise them at all, while others accept or reject just what pleases themselves."

Decades passed, and the nomenclatural muddle that Strickland had warned of only grew worse. Recognizing the problem but unable to agree on a way out, the world's zoologists fell into schism. German entomologists published one code of taxonomic nomenclature, the so-called Kiesenwetter Code. The American Ornithologists' Union published another. Lepidopterist W. A. Lewis and coleopterist D. Sharp each published their own code, as did van Maehrenthal and Banks and Caudell. Slowly, painfully, communicating via letters and telegrams and at an interminable series of international conferences, including the Berne Meeting, the Boston Meeting, the Graz Meeting, the Monaco Meeting, the Budapest Meeting, and the Lisbon Meeting, zoological taxonomists clawed their way into agreement, achieving unity by the first decade of the twentieth century, more than eighty years after Strickland first proposed his rules, and more than fifty since he died.

It is nearly unbearable to read a detailed history of these events, as I have done. Revolutions, which come into existence suddenly and catastrophically, make for interesting stories, but that is rarely the case with institutions. Vehicles to carry an endeavor across generations, bulwarks against the individual ego, institutions form by the slow, sedimentary accretion of collective effort. Given how difficult they are to build, people tend to treat existing institutions as immovable, neither changeable nor replicable. Conversely, people who don't like an institution's current condition often see it as fatally blighted, fit only to be torn down.

Today Strickland's institutional dream of rule and order is fulfilled in the form of several nomenclatural codes. One covers how taxonomists give names to microorganisms. Another covers the naming of plants, algae, and fungi. A third, called the International Code of

Zoological Nomenclature, covers the naming of animals. Now in its fourth edition, this zoological code runs to nearly one hundred pages and comprises eighty-nine articles and hundreds of subarticles and sub-subarticles. Most of the rules could be summed up as minutiae, dealing with grammar, spelling, and conjugation, but a few stand out. One is the principle that the Code pertains only to the act of naming, not to the taxonomy itself, "which must not be made subject to regulation or restraint." That is, the nomenclatural filing system must not impinge upon the science. Another important rule is the principle of priority: Assuming that it meets the Code's other criteria, the oldest scientific name given to a creature is the one that sticks.

∼

While Hugh Strickland argued that a Linnaean binomial was arbitrary, functioning the same way that the word "cow" immediately calls to mind a horned milk giver with four stomach compartments without actually describing any of those attributes, he thought that in practice, names should be at least somewhat connected to the creature to which they were affixed. But even this modest plea for description in naming is now cast aside. The Zoological Code allows that, so long as the grammar, conjugation, Latinization, and other procedural requirements are attended, any creature can be given nearly any name. A small plea for restraint comes in the Zoological Code's code of ethics, contained in the appendices. Compared to the Code itself, this code of ethics is austere, comprising just seven so-called principles. One is that zoologists should not describe and name a creature if they are aware that another scientist has already recognized and is working to name that creature—that is, thou shalt not steal. Another principle is that zoologists should not publish a name that "would be likely to give offense on any grounds." A third admonishes against the use of "intemperate language." But these principles are merely recommendations, points out Neal Evenhuis. "There's nothing that says you can't do it," he says. "It just says, you know, you really shouldn't."

Evenhuis is a taxonomist at the Bishop Museum in Hawaii who specializes in flies. He also sits on the International Commission on Zoological Nomenclature, the twenty-seven-member body that settles disputes involving the International Code of Zoological Nomenclature. He has named some seven hundred species of flies. He has named a fly *Phthiria relativitae*. He has named flies *Carmenelectra shehuggme* and *Carmenelectra shechisme*, after the *Baywatch* star. He has named flies after Charlie Chaplin and Milli Vanilli. He has named flies after Oreo cookies and jambalaya. He has named flies *Reissa roni*, *Riga toni*, *Pieza pi*, *Pieza rhea*, *Pieza kake*, and *Pieza deresistans*. He named a fly *Brachyanax thelestrephones*, which translates to "little chief nipple twister." "I like to do humorous things," he explains.

On his website, taxonomist Mark Isaak has amassed a sprawling collection of other silly names: a single-celled microflagellate named *Kamera lens*; a fish named *Stupidogobius*; a wasp named *Hakuna matata*; fungus beetles named *Gelae baen*, *Gelae donut*, *Gelae rol*, *Gelae fish*, *Gelae belle*; a wasp named *Humbert humberti*; a wasp named *Panama canalis*. In the 1850s, Reichenbach named an orchid genus *Aa*. In 1906, Semenov named a beetle genus *Aaata*. In 2013, Bellamy named another beetle genus *Aaaaba*. One group of moths is named *Eucosoma bobana*, *E. cocana*, *E. dodana*, and etcetera. Another group of moths is named *Hysterosia biscana*, *H. discana*, *H. riscana*, *H. viscana*, and etc. A group of plant hoppers is named *Cedusa bedusa*, *C. cedusa*, *C. gedusa*, &c. In the 1920s, Dybowski named a series of crustaceans *Crassocornoechinogammarus crassicornis*, *Parapallaseakytodermogammarus abyssalis*, *Zienkowiczikytodermogammarus zienkowiczi*, *Toxophthalmoechinogammarus toxophthalmus*, *Siemienkiewicz-iechinogammarus siemenkiewitschii*, *Rhodoph-thalmokytodermogammarus cinnamomeus*, and *Grammaracanthuskytodermogrammarus loricatobaicalensis*. A hundred years later, Chambers named a type of predatory soil bacteria *Myxococcus llanfairpwllgwyngyllgogerychwyrndr obwllllantysiliogogogochensis*, in reference to the Welsh parish of Llanfairpwllgwyngyllgogerychwyrndrobwllllantysiliogogogoch.

Evenhuis's *C. shechisme* echoes the prurient efforts of Kirkaldy, who gave a series of bug genera the names *Ochisme*, *Dolichisme*, *Florichisme*,

Marichisme, Nanichisme, Peggichisme, and *Polychisme.* Heinrich named a group of moths *Gretchena delicatana, G. dulciana, G. amatana,* and *G. concubitana,* meaning, Mark Isaak writes, "delicate," "sweet," "beloved," and "lying together." Whetzel named a gray mold *Botryotinia fuckeliana.* Fabricius named a wasp *Exetastes fornicator* and a beetle *Monochamus titillator.* Tilesius named a salp *Thetys vagina.* Linnaeus named a butterfly pea *Clitoria.* He named an earwig *Labia minor.* He named a stinkhorn fungus *Phallus impudicus.* He named one mollusk *Cypraecassis testiculus.* He named a second *Verpa penis.* He named a third *Volva volva volva.* In *The Naming of the Shrew,* John Wright notes the many binominal names ending in *anus.* He soberly explains that the Latin suffix "simply indicates position, connection, or possession by." Examples include *Ginkgoites garlikianus, Soranus,* and *Dolichuranus.*

Neal Evenhuis told me he views the act of naming as an opportunity to engage the public, bringing needed attention and, hopefully, funding. "Taxonomy, it doesn't have to be a dull, dry subject," he says. "You can make it interesting for people."

But sometimes the names that catch people's attention are not the ones that most taxonomists would want to be known for. Scientists have named species for Hitler, Mao, Lenin, and a host of more ordinary murderers and rapists, along with at least one pedophile. They have named species for impolite words and concepts, for racist and sexist slurs, for disgusting bodily functions. Most of all, they have named species for white men. In recent years, as social justice protests swept across the United States, Europe, and elsewhere, as famous and less-famous men were called publicly to account for their malignities, as people toppled statues and noted the moral shortcomings of John Muir and Charles Darwin and other former luminaries, taxonomists have found themselves defending these names, not on their merits but on their priority—that is, because they were there first.

The people they are defending these names from are, for the most part, nontaxonomists, who tend to view the principle of priority with less reverence. A few years ago, for instance, a pair of botanists proposed banning new botanical, mycological, and algological names derived from the word *caffer*, from the Arabic word for "infidel," now a highly offensive racial slur in South Africa. The amendment would also change the roughly three hundred existing such names by removing the *c*, leaving species names like *affer*, *afra*, and so on. Another pair of botanists proposed adding an article stipulating that a taxonomic name that was "culturally offensive or inappropriate, because it is (a) derogatory or insulting to a person or group of people, (b) is named in honour of a person that the taxonomic community agrees should not be honoured, or (c) otherwise causes deep offense," should be rejected. A third proposal suggested a series of amendments to the botanical code that would set aside the principle of priority's usual start date, which for botanists is 1753, the publication year of Linnaeus's *Species Plantarum*, in instances where an earlier Indigenous name for the same species could be found. Another group of scientists proposed not only that taxonomists end the practice of naming species after people but also that they rename species already thus named.

Even as taxonomists professed their personal commitment to social justice and morality in general, they responded to these proposals the same way again and again: The names could not be changed. Some raised practical concerns, noting that roughly a fifth of named species, or 375,000 species, give or take, are named after people, and that renaming them would set back the collective taxonomic task by decades. Others made philosophical arguments, calling the proposals a fascistic impingement on the freedom of scientific thought, a neo-colonialist imposition of Western social trends on the already overburdened taxonomists of the world's species-rich tropical countries, or an Orwellian erasure of the history of science. Still others pointed back to the nomenclatural filing system itself, the way a taxonomic name's role in denoting a category of creatures is theoretically separate from the meaning of that name, which is often fraught with cultural significance and reflective of the ego, prejudice, and other defects of the person who bestowed it. "Taxonomy is about naming species and other groups of organisms—animals, plants, fungi. So that is the science part," said Thomas Pape, a Danish blowfly taxonomist and current president of the International Commission on Zoological Nomenclature. "But in order to communicate, we need to have names." He continued. "What I think people are missing very often is that the whole idea of scientific naming is to have stable names. It's super nice if they are pretty, if they're easy to memorize, but the bottom line is stability." While nontaxonomists understandably often focus on the meaning of names, to taxonomists, the highest purpose of names is as headings in the nomenclatural filing system. Even bad names still served their purpose, he said—take *Anophthalmus hitleri*, the beetle named after Hitler. "Hitler did things that were, I mean, grotesque, unbelievable," Pape said. "But this is a name. I don't mind using that name."

Taxonomic nomenclature is the most unruly, outrageous, silliest filing system ever devised. It is the perfect companion to the science it is meant to support. Biological taxonomy, humanity's attempt to discern the underlying order of life on Earth, progresses in the manner of most sciences, by hypothesis and experiment, by the weighing of data and observation. But the results of these scientific efforts are unusually—perhaps uniquely—subject to the whim, notions, and proclivities of the people doing the weighing.

The trouble is evolution. While Linnaeus thought that at least some of the taxonomic rankings he proposed were real phenomena, created and maintained by "the Great Author," Darwin and Alfred Russell Wallace offered strong evidence that, via a process of natural selection, life-forms evolve over time. Biological taxonomy is therefore a series of rigid dichotomies—*genus* or *not-genus*, *family* or *not-family*—set upon a fluid reality. "Species," our basic unit for measuring life's diversity, is especially slippery. In *Origin of Species*, Darwin seemed impatient with the question, writing, "I look at the term species, as one arbitrarily given for the sake of convenience to a set of individuals closely resembling each other, and that it does not essentially differ from the term variety, which is given to less distinct and more fluctuating forms."

Scientists have since offered dozens of ways of defining "species," all of which serve reasonably well when comparing two obviously dissimilar creatures, like hippos and tigers. Where the definitions fall apart is at the edges, when it is less obvious where one species ends and the next begins. It happens that this liminal space is exactly where taxonomists spend much of their time. With no objective truth to guide them in the question of where to draw that line between *species* and *not-species*, they must rely on their own sense of things—a sense that, it turns out, is highly variable. Some taxonomists, popularly known as "lumpers," are inclined to see fewer species. Others, known as "splitters," are inclined to see more. Still others seem to see species nearly everywhere they look.

One such person was Constantine Samuel Rafinesque. Born in 1783 in Constantinople to a Greek German mother and French father, Rafinesque recalled his childhood as one of indiscriminate study—of natural history, philosophy, chemistry, medicine, horticulture, ancient Greek, and Latin. "I could have thus embraced any profession I chose," he wrote in his memoirs. Indeed, he worked variously as a law clerk, banker, professor, librarian, merchant, inventor, and brandy distiller. Late in life he wrote a 5,500-line epic poem titled *The World, or Instability* ("The fixed Polyps, under briny waters / Dwelling on rocks, can neither creep nor swim; / But yet they move, expand their limbs and feelers"). The profession he most often embraced, though, was taxonomy.

When he was eighteen, he traveled to the United States with his brother, arriving in Philadelphia on April 18, 1802. He immediately spotted a plant that was new to him. He thought, for some reason, that it was also new to everyone else. He called it *Draba americana*. "The American Botanists would not believe me," he wrote. He was undeterred. By the time he died, in 1840 (of cancer—induced, one medical historian proposed, by his long habit of treating minor ailments with a tea made of carcinogenic maidenhair ferns), he had described and named 6,700 new species and 2,700 new genera of plants. Other naturalists believed in hardly any of these names, which they said Rafinesque had affixed to creatures that other people had already described and named. "The passion for establishing new genera and species, appears to have become a complete monomania," one eminent botanist wrote soon after Rafinesque died. "This is the most charitable supposition we can entertain."

In other fields, scientists would be tempted to ignore the contributions of such a colleague. But taxonomists could not look away. If they wanted to name a new species in any of the groups that Rafinesque worked on, they needed to be sure that Rafinesque had not blundered onto an actual new species and put a name to it. The principle of priority and other rules of nomenclature, meant to provide stability and coherence, thereby became a set of shackles, chaining other taxonomists to the work of their least talented peers. "After years of effort devoted in part to consideration of the unending series of problems . . . raised by Rafinesque's work," wrote one botanist in the 1950s, "my frank conclusion is that in taxonomy and nomenclature we would have been infinitely better off today had Rafinesque never written or published anything appertaining to the subject." By one recent estimate, scientists now accept about fifty of Rafinesque's genera and perhaps three hundred of his species—less than half of 1 percent of what he named.

Another prolific taxonomist was Francis Walker. A friend described him as "a most disinterested benevolent man, highly educated, possessing great taste in the arts, himself a superior amateur artist, well versed in the sciences, and a perfect gentleman, in mind, manner, and conduct, throughout his life. He had never been in any business, and was untainted with any of its deteriorating effects." Thus unencumbered by the specter of financial responsibility, Walker instead nurtured a lifelong obsession with insects, particularly the tiny wasps that lay their eggs on or inside the bodies and eggs of other insects. In 1848, after an early career of some promise, Walker was hired by the director of the British Museum to survey the museum's insect collection. The resulting catalogue, which took him decades to complete, contained forty-six thousand species, almost sixteen thousand of which Walker described himself—a rate that one writer estimated was more than a new species a day.

In an obituary published in *The Entomologist's Monthly Magazine* in 1874, an anonymous writer summed up the quality of Walker's contributions: "More than twenty years too late for his scientific reputation, and after having done an amount of injury to entomology almost inconceivable in its immensity, Francis Walker has passed from among us." Another writer recalled, "He worked in a purely mechanical fashion, without grasp of the subject or principles of classification." A third noted, "He is responsible for no fewer than eleven synonyms of the well-known *Eutachina rustica* . . . the description in every case being based on a single specimen." Overall, scientists estimate that Walker described more than 23,500 new species. Many, if not most, were synonyms, species that had already been described and named.

A third famous taxonomist was André-Jean-Baptiste Robineau-Desvoidy, a French bachelor who loved flies. "Flies, you who have always given me my most cherished delights," he wrote. Modern observers note that in some regards Robineau-Desvoidy was ahead of his time, with forward-looking techniques and theories of classification. But he seems to have been bedeviled both by the poor quality of available scientific instruments and by his own innate lack of discrimination. "It is becoming impossible to define them clearly by means of words," he complained of one group of flies. "One sees the endless repetition of the same phrases and the same words. Their complete definition will push human patience to its absolute limits." His struggle for words reflected not only the paucity of the vernacular but also the lack of actual differences among the flies. Today, Robineau-Desvoidy is best known for having given 248 different names to what later taxonomists judged to be a single species of fly. "Dying in Paris in 1857, he was buried with his horse and his dog," a taxonomist who specialized in the same group of flies later wrote. "His face was singularly ugly."

Taxonomists, compelled to engage continually with the work of even their least talented peers and predecessors, have tried to understand what force compels the most fervent and indiscriminate among them. Francis Walker gave a small hint, admitting in a letter to a friend that he struggled at his mission but could not stop. "It has occurred to me that I am unequal to the task of describing these minute insects . . . with sufficient clearness," he wrote, "but from vanity & the pleasure of examining them I have been unable to desist."

This compulsion to describe and name, which blinded Walker, Rafinesque, and Robineau-Desvoidy to the creatures they were naming, seemed to their peers like some kind of disease. In the 1860s, a German taxonomist specializing in beetles accused a prolific colleague of being infected with *Mihisucht*. The first part of the construction comes from the Latin word for "to me" or "mine," which taxonomists once wrote next to newly described species to signify that they had coined a given name and were using it for the first time in print. *Sucht*, meanwhile, is a German word for "addiction" or "mania." Adopted into English, the disease became known as the mihi-itch.

~

This, then, is the story so far: Linnaeus drew up a blueprint for a great tower that could bring order to the chaos of life on Earth, and he offered its builders a common language, a way to transcend the chaos and babble of ten thousand folk taxonomies. But his disciples were indulgent of their egos, convinced of their own righteousness, and soon chaos threatened again, until Hugh Strickland and other like-minded lovers of order and rules convinced them all to bind themselves under a system of governance. Order was restored, mostly, but this nomenclatural institution had the side effect of leaving open in taxonomy a door that was long closed in other branches of science, whereby anyone, whatever their credentials or abilities, could enter, as long as they followed the rules. Chaos slipped back in, in the form of men captured by the need to name things, whether for glory or false principles or even a desire to sow confusion. The whole sequence has a strangely recursive quality, like a hoop snake rolling downhill.

It is here that this story of taxonomists begins to coil together with another, which at first had seemed entirely separate, even opposite. Where the story of taxonomy is about the scientific quest for order and understanding, this second story is about fear, fraud, and other facets of humanity's fallen nature. It is the story, in other words, of the snake men.

CHAPTER 4

SNAKE OIL

Then he rushed to the museum, found a scientific man—
"Trot me out a deadly serpent, just the deadliest you can;
I intend to let him bite me, all the risk I will endure,
Just to prove the sterling value of my wondrous snakebite
    cure."

—BANJO PATERSON, "JOHNSON'S ANTIDOTE"

The first famous Australian snake man was Charles Underwood. Colonial newspapers described him as "the celebrated snake charmer," "the notorious snake-charmer," "the stout able man," "bilious-looking," and "professor." Some said he learned his tricks in North America; others said he learned them from Amazonian shamans. Underwood's contemporaries were unclear on many of the details of his life, even before his villainous rival Joseph Shires began claiming that *he* was Charles Underwood. New South Wales, furthermore, was for a time home to at least two and perhaps three actual Charles Underwoods. Horatio, the middle name of *the* Charles Underwood, appears only in the record of his birth, on February 19, 1807, to Richard and Elisabeth Underwood of Pleasant Row in London, and in the report on the inquest into his untimely death.

The barest trail of crumbs traces Underwood's path through early adulthood. In March 1829, he was convicted in London on two counts of highway robbery and sentenced to fourteen years. At the time, this meant transportation to the penal colonies, which, whatever the purported length of the sentence, was most often a one-way trip. He is listed among the two hundred convicts who arrived in Sydney aboard the barque *Norfolk* in August 1829. His whereabouts for the next decade are uncertain, although later, as his fame grew, a Queensland newspaper wondered if he was the same Charles Underwood who had spent time in a penal colony there and who was known for keeping company with snakes. There is no further mention of him until 1841, when he was convicted of forging a check and again sentenced to transportation to a penal colony, this time from the Australian mainland to Van Diemen's Land, now known as Tasmania. When he reappeared again, in April 1849, it was as a snake man.

Underwood, newspapers record, had arranged a snakebite exhibition in a large room of St. Mary's Hospital in Hobart, Tasmania. On the appointed Saturday, about fifty learned men assembled to watch Underwood's experiments. Much was at stake: The lieutenant governor had authorized the presiding doctor to release Underwood from the terms of his bail if the experiments were successful. But difficulty piled upon difficulty. Despite running ads in the newspapers in the days leading up to the event, the learned men had procured only three snakes: a "large diamond snake," a "whip snake," and a "very small black snake." The snakes were too docile. When Underwood placed the provided rabbit in a cask with the snakes, they would not bite. Nor would they bite a cat. When Underwood took one of the snakes and forced its mouth to close over the rabbit's ear, it failed to envenomate the rabbit. Luckily for Underwood, during this last attempt, the snake bit him instead, finally giving him the opportunity to prove himself. He produced a "small phial containing a light-coloured liquid, and rubbed a portion over the bite," wrote one reporter. "He said it was sufficient to cure the bite of the deadliest species of snakes."

~

According to the ancient Greeks, a mighty serpent called Ophion incubated an enormous egg that hatched into the universe and all its inhabitants. Ophion also hatched "ophidiophobia," "ophidiophilia,"

Acanthophis, and other terms of serpentine twist. During his famous trials, Hercules encountered Echidna, who was half maiden and half viper. She had stolen his horses while he was napping. "I will give them back to you," she said, "but only after you have slept with me." The Gorgons had snakes for hair. Dante, who visited them in Hell, recorded that the trio, "who had the limbs of women and their ways," also wore girdles made of snakes. After Perseus cut off the head of Medusa, the most famous of the Gorgons, he carried it with him as he flew with his winged shoes over the deserts of Libya. Drops of Medusa's blood turned to snakes, which, Ovid explained, is "why Libya is now infested with poisonous reptiles."

During the Roman Civil War, Cato led a Republican army through the deserts of Libya, where his soldiers encountered heat, sand, and the various spawn of Medusa. As Lucan recounted in his *Pharsalia*, these included the chersydros, the chelydri, the cenchris, the cerastes, the scytale, the pareas, the dipsas, the drowsy asp, the deadly seps, the fierce haemorrhois, the dreadful amphisbaena, the swift jaculi, the scorching prester, and the basilisk. One soldier, named Aulus, stepped on a dipsas. It bit him, and soon his whole body was aflame and he was consumed by an unbearable thirst. A scorching prester bit another soldier, Nasidius, who immediately swelled to great size, bursting his coat of mail, until "he himself lay concealed, completely hidden within his swollen body." The unlucky nobleman Tullus was bitten by a fierce haemorrhois, causing blood to pour from his eyes, nose, ears, mouth, and other orifices and outlets. The soldier Sabellius was bitten by a deadly seps. His skin fell off, the membranes separating his organs dissolved, his bones turned to mush, and his head rolled away.

European colonists in Australia encountered a suite of venomous snakes just as fearsome as those of the Libyan desert. Australian newspapers in the early 1800s offered constant snake coverage, giving a sense of collective dread. There was a steady toll: women, men, "a fine boy," "a fine cow," "a fine ewe," all dead of snakebite. Some victims fell immediately into unconsciousness. Others vomited and lost their sight. One boy's tongue turned yellow and his teeth turned black.

Another man reenacted the death of Tullus: After he was bitten, blood gushed from his ears, eyes, nose, and mouth, and his body dissolved, becoming "instantaneously a mass of putrefaction, so that it was with difficulty removed into a grave."

The bite of a venomous snake is a mysterious thing. The wounds are small and often nearly painless in the moment. Sometimes death comes quickly, but other times it may not arrive for many hours or even days, during which time the victim will suffer any number of symptoms, including pain or numbness, paralysis or spasm, a pulse that races or drags, insensibility, confusion, terror. In their desperation to avoid this fate, people have turned to many different cures. Early remedies for snakebite included ether, tincture of asafoetida, mercury, strychnine, carbolic acid, chlorine of gold, chlorinated lime, mustard poultice, potassium permanganate, whiskey, brandy, gunpowder, petrol, toad urine, human urine, olive oil, wild radish oil, daffodil juice, rancid butter, goat cheese, goat milk, pig's lard, and pigface plant juice, as well as suction cups, ligature, and impromptu amputation.

Most of these methods worked roughly as well as modern products that promise to thicken the penis or restore memory and concentration to the addled. In one sense, snake men like Underwood were no different from present-day gadget and ointment salesmen. But the snake men drew from a deeper cultural well. In handling venomous snakes and weathering their bites, they were acting out an age-old story, of Hercules, Thor, Vishnu, Saint George, Saint Patrick, and other heroes who grappled with serpents, defeating evil, death, even fear itself.

~

To the astonishment and delight of the assembled learned men, Underwood seemed to suffer no ill effects from the snakebite—all thanks, he told them, to his proprietary cure. In March 1850, he conducted further experiments for men of science, this time successfully causing black, brown, and diamond snakes to bite two cats and twenty-three dogs, of which, one reporter wrote, "28½ per cent died after the application of the antidote, and 27¼ per cent without the antidote."

He began advertising his antidote in the newspapers. In 1852, he was sentenced to eighteen months of hard labor for stealing five gallons of rum. In 1854, reporters spotted him performing for large crowds in downtown Hobart. "He places the reptile's head in his mouth," one reporter wrote, "and allows it to twirl and wind about his person." A week later, one of the snakes bit Underwood on the tongue. A reporter wrote that "the charmer admitted he was under the influence of liquor." In 1855, a policeman discovered him lying drunk in the streets. When the policeman tried to arrest him, Underwood threatened him with a bag of snakes. A court later banished him from Hobart.

His growing fame seems to have inspired imitators. In 1857, Underwood published an advertisement cautioning that he could not be answerable for a cure purchased anywhere but at Mr. Millhouse's Manufactory, and that "no other establishment has the genuine article but him." In December 1858, newspapers reported the death of George Henwood, an itinerant clock repairman and snake charmer who sold what he called "Underwood's antidote." After he was bitten by one of his snakes while showing them off at a pub, Henwood "took a small bottle out of his pocket and rubbed some of the stuff that was in it on to his finger where it was bleeding," recalled one witness. "In ten minutes afterwards he came up into the bar; he looked very pale, and [the bartender] gave him a nip of gin, which [he] asked for; he then went into the kitchen and sat down; I asked him how he felt, and he said, 'better,' and then asked for some peppermint, which I gave him; he then went up stairs and sat in his bed; in about ten minutes I went up and asked him how he was; he said, 'better.'" Henwood died soon after. In an article on the case published the following February in *The Australian Medical Journal*, E. Swarbreck Hall, Esq., MBOS, wrote that "it is very necessary to warn the public from reposing confidence in mere external antidotes . . . as was unfortunately the case, from this snake-charmer's trust in Underwood's antidote."

In April of that year, Underwood appeared in Melbourne and announced a snake exhibition. "He does not state whether he will permit the snakes to operate upon his person," the newspaper notice said, "but persons are politely requested to bring their own snakes." In May, *The Cornwall Chronicle* revealed the source of Underwood's power. In a Rumpelstiltskinian turn, the newspaper reported, Underwood had appeared late one night at a shepherd's hut and asked to borrow a pot. The shepherd observed as Underwood boiled and stirred and filtered his brew. After the snake man left in the morning, the shepherd scuttled down to the newspaper offices. "The plant that possesses the quality of neutralizing the effect of the venom imparted to the human body by the bite of the snake," the paper declared, "is no other than the common fern!"

In a letter published soon after in *The Argus* newspaper, Underwood wrote, "I defy all the medical powers in the world to save the life of any person or animal bitten by snakes with fern leaves." He then offered to tell his secret "for the small sum of one penny per head for the population of Victoria and its dependencies." Another newspaper calculated the sum to be roughly 3,000 pounds, or something near a half-million dollars today.

~

It was around that time that Underwood seems to have caught wind of his most brazen imitator. The first record of this feud is from late October 1859, when *The Bendigo Advertiser* reported that Underwood had appeared two nights before at the Victoria Hotel in Bendigo and allowed himself to be bitten on the cheek by a snake "kindly lent" by one Dr. Hutchinson. "The effect of the bite was very soon perceptible," the reporter wrote. "A stupor appeared to be commencing, the pupil of the eye became dilated and insensible to light, and a slight rigidity of the limbs apparent." He applied his antidote and soon recovered. During his short convalescence, though, a member of the audience declared that this Charles Underwood "could not be *the* Underwood." The meeting fell into argument, until Dr. Hutchinson took the stage and declared that, whether or not this Underwood was *the* Underwood, he had in fact been bitten by a venomous snake, applied his antidote, and survived, thereby fulfilling all expectations of how a Charles Underwood ought to behave. The *Bendigo Advertiser* reporter then quoted from an article that had run the day before in the newspaper in Castlemaine, a town about twenty-five miles away: "A gentleman of the highest respectability showed us a letter from Melbourne yesterday, in which it is stated by a competent authority, that the individual who is now experimenting with snakes in Castlemaine is not the real Underwood, but a person named Joseph Shires." The *Bendigo Advertiser* reporter demurred from judgment, writing that "as to the allegation conveyed we cannot offer any opinion," but noted that the man now claiming to be Charles Underwood at the Victoria Hotel had a tattoo on his arm that read "Joseph Shires."

Shires's beginnings are as mysterious as Underwood's. He was born under a slightly different name, Joseph Beaumont Shiers, around 1819 in New York, New York. In the file on his 1841 conviction in London, for stealing a coat, watches, a ring, and other goods, he was described as possessing a sallow complexion, an oval head, a long visage, a small mouth, brown hair, black eyebrows, blue eyes, and no whiskers. "Protestant," the document noted. "Can read and write." Sentenced to seven years' transportation, he arrived in Tasmania the following year, among 250 convicts aboard the *Surrey*. In the public notice announcing the ship's passengers, Joseph Shiers was transformed into Joseph Shires—perhaps a clerical error, perhaps a hint of his tendency toward reinvention. Over the next two decades, records are few, most of them noting criminal convictions for fighting in bars, stealing a watch, deserting a whaling ship, and beating his wife. The first time his name is mentioned in connection with snakes is in an advertisement he placed in the Melbourne *Age* in August 1859, two months before the Bendigo exhibition, inviting people to watch him test Underwood's antidote with two snakes that he'd caught. He signed the advertisement with his own name, apparently only deciding later to adopt Underwood's.

Upon hearing of the Bendigo exhibition, the real Charles

Underwood offered the imposter a public challenge: They would duel by snakebite, and see whose cure was better. Shires accepted the duel, and in late January 1860, the two men met at the Cornwall Assembly Room in Launceston, Tasmania. Advertisements for the free event drew what a reporter deemed "a motley assemblage." Underwood appeared first. He was dressed in a blue coat, black pants, and a black hat, and wore "a peculiar smile." When he lifted his hat, a black snake tumbled out. He began charming it, guiding it back and forth on the floor while murmuring incantations to the audience. He decried Shires the imposter, and the audience began to chant "Shires, Shires!" Shires appeared. He wore no coat or hat, and the tattoo on his arm was visible. Underwood and Shires shook the long, forceful handshake of bitter enemies. They shouted over each other for the audience's attention. Underwood put the snake back in his hat and took it out again. Shires put the head of a snake in his mouth, then said it bit him, as blood trickled from his lip, though the reporter wrote, "We cannot say whether he did not himself inflict the wound with his own teeth." The snake charmers tried to test their antidotes on kittens, but the snakes refused to bite. "The exhibition was absurd from beginning to end," the reporter concluded, "and utterly valueless in settling the question, no doubt a very important one to all, whether or no there is an antidote for snake bites."

Several months later, another advertisement appeared on the inside pages of *The Argus* newspaper:

ANTIDOTE for SNAKE-BITES.—Challenge to Charles Underwood.— The undersigned CHALLENGES you to MEET HIM, on Monday evening next, the 30th of April, at half-past seven, at Mr. Taylor's, All Nations Hotel, Sandridge, there and then to TEST our ANTI-DOTES for SNAKE-BITES. The arrangement to be, you bring snakes by which I will be bitten and apply my own antidote, and, vice versa, you to be bitten by snakes I produce, applying your antidote.

JOSEPH SHIRES

The advertisement makes no mention of their prior encounter, nor do the two men appear to have mentioned it during the resulting duel. They met a third time two weeks after that, at the Union Hotel in Geelong, southwest of Melbourne. Witnesses said both men appeared to be drunk. None of their contemporaries seems to have realized what is now obvious: The two snake men were working together. The duels were a sham, a way to sell snakebite antidote. The conspiracy, though, was short lived. Underwood died a little more than a year later, in November 1861, after one of his snakes bit him while he was showing it off in a bar. In the report on the inquest into events leading up to his death, the landlady of the Eureka Hotel recalled that Underwood and his wife and young daughter entered the bar "about half-past seven o'clock on Saturday night. They had among them a cage of snakes. Could not say whether [Underwood] was sober or not, but should not think he was."

Joseph Shires was now Australia's most famous snake man. In the years that followed, he toured widely. Advertisements for his snakebite antidote appeared regularly in newspapers. For a time, it seemed that his snakebite-based fame and fortune could only continue to grow. Then he encountered Professor George Britton Halford.

An English anatomist with powerful muttonchops, Halford arrived in Melbourne in 1862, and soon joined the colonial quest for a snakebite cure. In early 1867, he published a description of his investigations into the physiological effects of snakebite. Using a microscope, he had discovered that snake venom suffused the blood of the bite victim with what he called "germinal" cells, which rapidly multiplied, preventing the blood cells from carrying oxygen, and causing "coldness, sleepiness, insensibility, slow breathing, and death." In a subsequent lecture, Halford noted that snake venom's effect on the blood was like that of cholera, and modestly suggested that the venom of dead snakes, dried and carried aloft on the wind, might be the cause of that disease. People wrote letters to the editor praising him. Colleagues lauded him. One Dr. Barker "was of opinion that if the inquiry is pursued further it will alter the whole system of pathology." It seemed Halford was on the brink of a major breakthrough.

In November 1867, apparently in the spirit of earnest inquiry, the professor publicly invited Shires to give an exhibition in Melbourne. Shires accepted, and in early December, they met. Halford was impressed. "I caused him to be bitten by a tiger snake on Monday morning," the professor wrote in a letter to the editor published that Thursday, December 5. He then applied a few drops of Shires's antidote to the wound. "The arm soon began to swell, but with the exception of nausea, sickness, and some diarrhea, and restlessness, followed by drowsiness, no evil results have followed," he continued. "There is no deception about this man."

But less than three weeks later, Halford published another letter to the editor, this time describing the results of a subsequent experiment with Shires's antidote. He had induced tiger snakes to bite nineteen dogs. He treated twelve of the dogs with Shires's antidote and did not treat the other seven. One of the dogs he treated survived. So did one of the dogs he didn't treat. He induced a tiger snake to bite this last dog again. It died. "As regards the value of the antidote as a means of saving dogs," he concluded, "your readers are now as good judges as myself."

Halford seems to have realized that his association with the snake man was costing him in the public eye. Some newspapers pointed out that the medical doctor should not have "caused" Shires to be bitten, and that if Shires had died, he would have faced manslaughter or murder charges. Scientific skeptics, meanwhile, speculated that his germinal cells were in fact ordinary white blood cells. *Melbourne Punch* published a mocking letter in which one H.B. Halford A.R.T.F.U.L. &c. described a series of snakebite experiments. "No. 2—A young donkey was bitten by my largest carpet snake. Went into convulsions, turned black in the face, said '*Oh, my poor mother!*'" In January 1868, Halford wrote a letter to *The Argus* clarifying that when he said "there is no deception about this man," he meant only that Shires had indeed been bitten by a snake.

In his own letters to the editor, Shires protested Halford's slander. "He must admit it was my antidote alone that saved him from being accessory to my death," he wrote in one. "Until the doctor can produce a more certain remedy . . . he ought the rather to encourage the use of mine." In another letter, he wondered, "Why cannot I parade my cures before the public without drawing down upon me and my living a term usually applied to those who wish to practice deception and deceit?" A few months later, his reputation suffered a further blow when he agreed to let one of his tiger snakes bite a police magistrate who did not believe that the snakes were venomous. Shires was eventually acquitted of manslaughter and began performing again, but he never regained his former status.

Halford, meanwhile, continued his quest for a snakebite antidote. In November 1868, just as media coverage of Shires and the dead police magistrate faded from the papers, he announced in a letter to *The Argus* that he'd discovered the cure: ammonia, injected straight into the veins. "This mode of treatment need not be limited to snake-poisoning," he wrote, "but might, perhaps, be extended to opium-poisoning, or to that resulting from infection, as in fever, cholera, &c." Halford did not mention in the letter how he'd hit upon the idea, but months before, newspapers had quoted him saying that he believed Shires's proprietary antidote was mostly ammonia. His new antidote, which was perhaps just Shires's old antidote, was an immediate success. Newspapers filled with reports of snakebite victims miraculously cured. Within weeks, advertisements for ammonia-injecting kits appeared. "Important to Squatters, Farmers, and persons residing in the Bush," the advertisements declared. "So simple that anyone can easily use them." The device itself became known as the Halford syringe.

But just a few months later, in February 1869, a young woman in Sydney was bitten by a snake and died despite receiving ammonia treatment with a Halford syringe. Scattered murmurs that people ought not inject themselves with ammonia turned into a cacophony. Gerard Krefft, alleged pornographer and expedition mate of Blandowski, was among those who wrote letters to the editor casting aspersions on Halford's method of treatment. Before long, a panel of medical men invited Halford to participate in experiments in which they would cause snakes to bite dogs, then attempt to cure them by his method. The dogs all died, an even worse result than when he had performed the same experiment with Joseph Shires. "It cannot be said that the ammonia treatment was in any way beneficial," the panel concluded. In a letter published soon after, Halford insisted that "injection of ammonia into the veins is the speediest and most certain method" of preventing death by snakebite. The man of science had wriggled from his skin and turned into a snake man.

In 1891, the papers announced that Joseph Shires was preparing his will, said to include the recipe for his famous snakebite antidote, which, the papers noted, was "never operative in other hands than his own." A few months later, he died, ending as mysteriously as he had begun: The notice of his death said that the cause was a "tree falling on him." His recipe was never published.

By then, the Australian colonies were thick with snake men. Most of them followed Underwood and Shires's proven model, inducing venomous snakes to bite them and then curing themselves using their proprietary snakebite antidotes, which they then offered for sale to witnesses. To judge by the volume of associated newspaper articles and advertisements, these entrepreneurs were remarkably successful. Shires turned out to be unusual, though, for having died in a manner unrelated to his craft.

On May 3, 1893, snake man Victor Hullar was bitten on the finger by a tiger snake while performing in northern Victoria. He drank his proprietary snakebite antidote with some brandy, but soon after began to shake and fell unconscious. After a visit from a doctor, Hullar regained his senses, and for a time, his condition seemed to improve. "When the poor fellow gets well again," *The Bendigo Independent* suggested, "it will be an excellent advertisement for popularizing his show." But Hullar did not get well.

Exactly ten years later, on May 3, 1903, snake man Professor W. Pegleg Davis was bitten on the arm by one of his tiger snakes during a performance in Launceston, Tasmania. He applied his snakebite antidote to the bite, then finished the show. A friend who accompanied him home later recalled that Davis seemed drowsy but refused medical treatment. "I will be all right in a minute," he said. When a friend visited in the morning, Davis asked him to go get brandy. By the time he returned, Davis had died.

Another snake man of academic title was Professor Fred Fox, who in 1897 set up among the jugglers, trick shooters, puppeteers, buskers, and mountebanks who gathered at La Perouse, the last stop of the Sydney streetcar system. He performed there regularly, selling his proprietary snakebite ointment, until 1913, when he left his understudy, Garnett See, in charge of the show and set off for India, seeking to collect a £25,000 reward from the Indian government for the development of a snakebite antidote. On See's first turn as ringmaster, he was bitten by a brown snake and died. On the same day, in Melbourne, snake man Harry Deline was also bitten by a brown snake and died. A newspaper noted that "this was the second time within a week and the third within a few weeks that Deline had been bitten by a snake." Professor Fox met the same fate several months later, after inducing an Indian krait to bite him five times, and despite having smeared himself with snakebite ointment.

In 1917, a snake-man trio composed of announcer That Man Gray and performers George Valves and Barnett Albarez dwindled when Albarez was bitten by a tiger snake during a performance and died; four weeks later Valves was also bitten by a tiger snake, resulting in what the coroner described as "merely a scratch." That Man Gray continued to sell his famous Tiger Salve, known as "KING of all OINTMENTS," containing "90 per cent. more medicinal power than any other salve or ointment on the market," for several more years before going mad and dying in an asylum.

On March 13, 1920, Cleopatra Caton was bitten by a tiger snake while performing and died. Caton was one of the snake women, alongside Señorita Amoretti, Cleopatra Cann of La Perouse, Navada the Snake Girl, the winsome Princess Indita, Melinda Lee the Tennessee Snake Dancer, Paula Perry the Queen of the Snakes and Australian Bush Girl, Latiefa the Reptile Queen and Clever Young American Cowgirl, and Belle Wade, aka "Monica Sans" aka "Venitia" aka "La Belle." There were always fewer snake women than snake men in Australia, and they tended to be convicted of fewer crimes and were not involved, for the most part, in the purposeful envenomation of dogs and cats and themselves. Compared to the snake men, who offered displays of heroism and masculinity, the snake women hewed more to the beauty-and-the-beast model, with performances often bordering on striptease. This image remains intact into the modern era, exemplified by Britney Spears's 2001 MTV *Video Music Awards* performance of "I'm A Slave 4 U," in which she wore a bikini and an albino Burmese python. In 2014, the appearance of another snake at the *VMAs* was canceled after the snake, a boa named Rocky, bit a backup dancer during rehearsal for Nicki Minaj's performance of "Anaconda," which makes explicit the Freudian insight that snakes are symbolically and also morphologically phallic.

On July 23, 1921, Australian snake man Professor Tom Morrissey, né Thomas Joseph Wanless, was bitten by a green mamba while performing during a trip to Durban, South Africa. One newspaper reported that it was his "tenth bite in South Africa." Another recorded that the morning after the bite, following a difficult night, Morrissey looked himself in the mirror and calmly said, "The green mamba wins." Indeed it had.

Six months later, a snake bit Professor Morrissey's colleague, Anthony "Psycho" Kimbel, on the thumb while he was performing at a carnival. He died, too.

On April 23, 1922, snake man John Miller was bitten by a tiger snake while performing in Cohuna, Victoria. He refused medical aid, "professing confidence in a remedy of his own." His confidence was misplaced.

On May 19, 1927, C. J. French, curator of the snake park at the

Adelaide Zoo, was bitten by a black snake while performing for a group of children. "Assuring the crowd that there was no cause for alarm," *The Daily Mercury* reported, "the curator first applied a ligature, then cut and sucked the wound, and rubbed in some permanganate crystals." He proceeded with the demonstration and died that night.

On January 12, 1928, Dot "Cleopatra" Vane was bitten on the thumb by a tiger snake while performing at an amusement park in Perth. Despite receiving both hospital care and her husband Rocky Vane's proprietary herb-based antidote, she died.

On February 23, 1929, snake man Harry Melrose was bitten by a tiger snake while performing for the local chapter of the Royal Antediluvian Order of Buffaloes. "Got me that time," he said, according to witnesses, and died soon after. Questioned later by a judge, a witness said that Melrose was "not drunk at the time, though he had been drinking."

On April 6, 1931, snake man James Murray, who "expressed extraordinary confidence in his power to overcome the power of snakebite," who "handled the reptiles with rare skill and absolute fearlessness," who claimed to have "caught 93 snakes" in one day, was bitten by a tiger snake while performing and died.

On October 2, 1932, snake man Cecil O'Sullivan was bitten on the finger by a black snake while performing in Euramo, Queensland. Shortly before he died, he told a witness that someone had stolen the last bottle of his never-fail snakebite antidote.

On November 4, 1932, snake man John Greaves, also known as "Professor Davis," was bitten by a "large pet snake" while preparing for a snake show and died, despite applying his snakebite antidote to the wound.

On December 7, 1934, Julius Mitchell, also known as "Milo the Snake Man," also known as "Captain Gus Leighton from the Interior of South Africa," was bitten during a show at Kurri Kurri while sticking a tiger snake back in its box. He applied his antidote to the wound but continued to look sickly. "A doctor saw him and told him to go to hospital, but he said he would be all right," one newspaper reported. He was not all right.

~

Even as hordes of snake men wandered Australia's streets and byways, giving medicine shows and getting drunk and killing cats and dogs and themselves, a more sober effort was underway. In 1887, Henry Sewell, a physiologist at the University of Michigan, published a paper describing his experiments with rattlesnake venom. He had diluted the venom in glycerin, then injected pigeons with small doses of the mix. After killing a number of pigeons, Sewell arrived at a sublethal dose. Once this set of pigeons had recovered from the initial dose, he found, they were able to weather slightly larger doses; after that, even larger doses. The pigeons eventually became truly heroic, able to bear more than seven times the original lethal dose. A few years later, Albert Calmette, a French scientist and head of the Pasteur Institute's Saigon branch, mirrored Sewell's methods to immunize horses against cobra venom. Once the horses were sufficiently immune, Calmette drew their blood and separated out the serum (the clear liquid that oozes from scrapes), which he then injected into rabbits. This, he found, made the rabbits immune to cobra venom.

The Pasteur Institute quickly began mass-producing the "serum antivenimeux," which Calmette thought would counteract the venom of all types of snakes. But Charles James Martin, a medical professor at the University of Melbourne, found that Calmette's antivenom offered no protection from the venom of Australia's red-bellied black snake or tiger snake. In 1898, New South Wales medical officer Frank Tidswell injected a horse with tiger snake venom, eventually producing an effective antivenom. He injected rabbits with the antivenom, then with lethal doses of the venoms of tiger snakes, black snakes, brown snakes, and death adders. The antivenom reliably protected against tiger snake venom but, as with Calmette's antivenom, not against the venom of the other snakes. "It is obvious, therefore, that, although an effective serum has been obtained, its action is specific, being operative only against the particular kind of venom used in its production," Tidswell wrote. The antivenom, in other words, must match the venom.

This, not ferns or ammonia or toad urine or any other proprietary substance, is an explanation for the snake man's survival—and often his death. A person stands a decent chance of surviving any given bite, even from one of Australia's more venomous snakes. Having survived it, they stand an even better chance of surviving the next bite. But resistance to the venom of one species of snake offers little or no resistance to the venom of another. This explains why Professors Fred Fox and Tom Morrissey, who both reliably weathered the bites of Australian snakes, were undone by an Indian krait and a South African mamba, respectively. Other snake men may have died due to similar changes in their act, or because they had taken too many bites in too short a stretch, or because they skipped breakfast. Any livelihood based on being bitten by venomous snakes is surely tempting to the gods.

In the moment of being bitten, then, a person suddenly faces a taxonomic question, millions of years of mammalian fear and millennia of human myth-making all boiling down to this: What just bit me?

CHAPTER 5

SONG OF THE SNAKE

The path to my fixed purpose is laid with iron rails, whereon my soul is grooved to run. Over unsounded gorges, through the rifled hearts of mountains, under torrents' beds, unerringly I rush! Naught's an obstacle, naught's an angle to the iron way!

—MELVILLE, *MOBY DICK*

In the 1970s, when Kevin Markwell was a boy, his neighbors would sometimes kill snakes and leave them draped over the garden fence, like criminals hung up on gibbets as a warning to others. Markwell would take the bodies down and preserve them in jars of denatured alcohol. He understood, he later wrote, "that reptiles were undesirable and that my interest in, and passion for them, was unusual and a bit quirky." It was another mark against him—he had no interest in girls, and, almost as unusual among Australian boys, no interest in sports. But his sense of alienation was tempered by the presence of a hero who viewed the world as he did: the famous snake man, Eric Worrell.

"Just as other boys idolised rugby league and cricket players, I idolised him," Markwell wrote. He pored over Worrell's best-selling books and read his magazine articles, including "Snake Bite Need Not Be Fatal" and "Be Your Own Taxidermist," which Worrell published under both his given name and pseudonyms such as Sap-Engro, Karliboodi, and Belvedere. He watched Worrell's documentaries and TV appearances and collected newspaper clippings about him into scrapbooks. The idea of seeing him in person was nearly overwhelming. When his family would pull off the highway into the parking lot of the Australian Reptile Park, Worrell's most famous creation, Markwell wrote, "my belly filled with butterflies."

Most Australians seemed to agree with Markwell's neighbors that "the only good snake is a dead snake," he wrote. But inside the park there existed a different reality, "one in which reptiles were displayed, admired and celebrated.... Reptiles were exhibited in large open cement walled ... 'pits'. What appeared to be dozens of red bellied blacks, tigersnakes and brownsnakes sunned themselves in tangled piles a metre or so from my body." When we spoke on the phone, Markwell told me that for him, visiting the park "was like going to an art gallery and seeing Picassos and Rembrandts. You would see these iconic species of reptiles that you would not see anywhere else." There, kids like him were not quirky or unusual, he said. "We had a place."

Sometimes there in one of the pits, hoisting a writhing serpent, was Worrell himself. When he was two years old, Worrell won a beautiful baby contest, and he was correspondingly handsome when he was older, with a puff of curly hair, deep-set brown eyes, a large, aquiline nose, and a robust beard. He was bright and open, a modern kind of snake man. While he shared some characteristics with the snake men who preceded him, including alcoholism and the sometimes loose association with the truth that characterizes great showmen and salespersons and charismatic people in general, he elevated the form beyond its hucksterish roots, making it presentable to a broader public, even something to aspire to. He brought a revolutionary new outlook to the ancient profession, one based not on the implicit fear of snakes, a performance of ophidiophobia conquered, but on the converse, on a dispelling of ancient myths and terrors. He was a conservationist: People should respect snakes, he argued, not kill them.

This message is now so deeply ingrained in the modern snake man that it is almost inconceivable that it should be otherwise. But this is merely evidence of the depth of Worrell's influence. His interest in the latest science and his pioneering forays into the arcane and highly specialized realm of taxonomy, meanwhile, begins to explain why, out of all the people working to understand life's many regions, it was Australian snake men who were first accused of that strange and novel kind of crime, taxonomic vandalism.

Worrell was someone to emulate. Kevin Markwell owned a book titled *Australian Snake Man: The Story of Eric Worrell,* part of a series of kids' books about eminent Australian scientists. In the book, Markwell says, Worrell was portrayed as having grand adventures, all involving reptiles, none involving sports or female romantic interests. This vision of Worrell's life seemed to offer Markwell his own potential route. "I had no interest in going to university," he says. "I saw myself getting a job in a zoo. And obviously, the reptile park is where I really wanted to work. So he kind of gave me a template for my own life."

Over time, though, Markwell's dreams changed, as dreams usually do. He finished school and became a professor of leisure and tourism studies at the University of Newcastle. Nothing in his professional life involved reptiles, until one day, he says, "I just woke up and thought, 'I could write an academic article about the reptile park as a tourist attraction.'" The article evolved into the book *Snake-Bitten,* a biography of Eric Worrell, coauthored by Markwell's colleague, historian Nancy Cushing. The project offered him a chance to look back on the snake man through a more critical lens. Worrell had been "this hero who you project all kinds of things onto," he told me. "And then you realize that that hero is flawed."

~

One Sunday afternoon, the snake man of La Perouse noticed a little boy standing beside the pit. The boy had curly hair and wore shorts and a blazer, and he stayed for the whole show. The snake man, George Cann, was reminded of himself. When he was a little boy, he stood beside the same pit on many Sunday afternoons, watching Professor Fred Fox, the first snake man of La Perouse. It was there Cann met Snakey George, a hermit who lived in a shack in the bush not far from the pit and made a living catching snakes, which he sold to zoos, museums, and other snake men. Snakey George took Cann on as a kind of apprentice, and when he was twelve, Cann started performing snake shows. When he was sixteen, he watched Professor Fox's understudy, Garnett See, take a fatal bite during See's first turn as the snake man of La Perouse. After See's death, the Charlie Hessells, Sr. and Jr., took over Professor Fox's snake pit, sometimes in partnership with Tom Wanless, whose encounter with the green mamba in South Africa was still several years off. George Cann spent the last part of World War I in Europe and, on his return, became the newest snake man of La Perouse.

Cann was a better snake man than Garnett See and Tom Wanless, luckier at least, and he was still standing in the mid-1930s, when the little curly-haired boy appeared. The boy was there the next Sunday, and the Sunday after that. One day after the show, Cann let the boy help him carry a bag of snakes down the road to Cann's house. Soon he began teaching the boy how to handle snakes, just as Snakey George had taught him. Eventually Cann gave the boy a snake of his own. As Eric Worrell later recounted, "The day George Cann himself actually gave me a whip snake was one I boasted of for many months."

The whip snake joined a large menagerie. Worrell's parents possessed a spirit of acquiescence, allowing him to keep chickens, rabbits, guinea pigs, mice, a tortoise, a stray dingo, goldfish, lizards, and one snake, then many. While Worrell loved animals of all kinds, Kevin Markwell says that he seems to have focused on reptiles by the time he was ten or eleven. "But he does that kind of pragmatically," Markwell says. "He kind of looks around and sees that, well, there are people working on birds, there are people working on mammals, but there's no one really doing very much on reptiles."

Worrell himself recounted the beginnings of his life's work in more romantic terms. "Everybody has a dream, a goal they set themselves and either modify to suit circumstances as years pass, or discard when they grow too old," he wrote. "My dream has remained unchanged from the moment I became aware of my inclinations. . . . I wanted to live somewhere away from the city, on a big piece of land where I could keep all the strange reptiles I would bring back from my travels to all corners of Australia. I wanted to keep them in as near to natural condition as possible so I could study their life histories, and so that people could look at them and learn that there was no reason to fear them."

He left school in his early teens and began working a series of odd jobs, including picking fruit, mining gold, and collecting turtles, which he sold to Sydney pet shops. In 1939, when Australia entered World War II, Worrell joined a construction crew digging sewer lines near the Murray River. He was pleased to find the region rotten with tiger snakes. The crew was later reassigned to Queensland, in the country's far north, where locals soon came to know him as a snake man. They gave him ten-foot pythons as gifts and brought him strange creatures to identify. "Often I returned to my hut after a day's absence to find some mysterious or evil-smelling packages awaiting me," he wrote.

After the war, he worked in the Northern Territory as a collector of fish, snakes, and lizards for zoos, museums, and pet stores in the south, as a tour guide, and as a journalist, contributing articles to *Wild Life, Walkabout, Outdoors and Fishing,* and other magazines. He wrote on a wide variety of subjects, including crocodile hunting, the tropical rainy season, and ferreting for rabbits. ("Good, easily handled ferrets are essential.") Most of all, he wrote about snakes. In these articles, Worrell bridged the long-standing divide in the world of Australian snake professionals: on one side, the snake men who played to the emotional and cultural power of serpents, performing the ancient rite of human mastery over nature; on the other side the men of science, sober and dispassionate, interested in taxonomy and nomenclature, disdainful of the "antidote vendors" and "snake charmers" peddling "fables" and "canards." Worrell drew from both, spinning tales of drama and daring from his own snake-catching adventures as well as those of George Cann and other snake men, while delighting equally in offering factual tidbits and dispelling snake-related myths. Snakes "do not milk cows," he wrote in one article. "There are no such things as hoop snakes that place their tails in their mouths and bowl down hills. Snakes that crack themselves like whips do not exist."

Worrell's world was richly inhabited, with every creature, every bit of foliage, not just identified but characterized—"white-throated ta-ta lizards," "blue-winged kookaburras," "crimson blood-finches," "shivering umbrella grasses," "stout Leichhardt-trees." In the tradition of John Muir, he often contrasted the sublimity of nature with the cravenness of humankind. "Native passion-vines drape in wild profusion over rotted stumps and logs shattered by bombing," he wrote in one postwar article about the Northern Territory. "These wildly growing vines smother all immobile objects—camouflaging atrocities of man and adding to the natural beauty."

The message that he would deliver for the rest of his life was taking form. "Rivers were shot and reshot time and time again," he wrote in one article. "Then suddenly there were no more crocodiles." In another article, he wrote, "In some, almost inaccessible parts of the river we found tiger snakes in large numbers, but anywhere in reach of a township was hardly worth the search. It appeared from inquiries we made that a local sport was to take a car or motor bike for a weekend's snake shooting." In still another essay, he recounted a conversation he'd had with the ornithologist Dom Serventy:

One morning on Fisher Island research station I was awakened suddenly by Dr. Serventy stoking up the fire.

"Damn you, doc," I growled. "Couldn't you have waited a few more minutes? You've just spoilt my dream."

"I'm sorry, Mr Worrell, you should have told me. I apologize for my lack of consideration. Is it too late to make amends?"

"I'd have to finish the dream to know," I said. "I was dreaming that a fellow was showing off with what he claimed was a deadly, Chappell Island tiger snake around his neck. I took the snake from him and found that he had cruelly mutilated the jaws to remove the fangs and venom sacs. I was so upset that I sought you out, doctor, showed you the snake, and asked if you could help me to do something about it. Before you could do anything your fire woke me. . . . Now what would Freud have to say about that dream?"

Doctor pursed his lips and dramatically waggled his finger.

"I'm glad you asked me that question, Mr Worrell. It is all very clear. Last night before retiring you will remember expressing concern that insufficient was being done in Australia to conserve wild life. As you are a reptile man the snake was your medium. The snake and its mutilation symbolized fauna and its destruction. Subconsciously you were appealing to me, as a government official, to do something about it."

When Charles Darwin visited New South Wales in 1836, he wrote that "a few years ago since this country abounded with wild animals; now the Emu is banished to a long distance, and the Kangaroo is become scarce. . . . It may be a long time before these animals are altogether exterminated but their doom is fixed."

While the continent's strange life-forms entranced and confounded naturalists like Darwin, most colonists were less impressed. Here, the order of nature was strangely perverted, like the Antipodeans of medieval bestiaries, who had backward feet and faces on their butts. The birds had no song, the flowers no scent, the animals no flavor. Most Australian mammals carried their young in pouches. Others laid eggs. Trees kept their leaves and shed their bark. The landscape greened in the winter and lay brown and dormant through the summer. The swans were black. "The animals indigenous to Australia are rather insignificant," wrote one newspaper editor in the 1850s, expressing a common sentiment, "furnishing a little sport, an occasional meal, and an interesting study to the physiologist, but of little value for the practical purpose of every-day life."

In the century that followed the arrival of the First Fleet, colonists worked hard to remediate Australia's floral and faunal deficiencies.

They brought sheep, cattle, chickens, red foxes, alpacas, monkeys, blackbirds, thrushes, goldfinches, pheasants, starlings, and sparrows. They planted blackberries and prickly pears that spread beyond control. They released rabbits, which swelled to a plague that drove farmers from their land. "When the sun is hot you can go along the fences or any place where it is shady and kill hundreds with a stick," recalled one farmer, quoted by Eric C. Rolls in his 1969 book *They All Ran Wild*. "The paddocks stink with the dead ones." The colonists killed dingoes, Tasmanian tigers, and other creatures that might eat sheep. They slaughtered kangaroos, wallabies, and emus that might compete with sheep for grass. They shot kookaburras out of misplaced fear that the birds killed chickens. They scattered poison and set traps. Often, they seemed to kill only for the thrill of it, reveling in "the freedom of the bush, unshackled by the trammels of the British Game Laws," as Horace William Wheelwright put it.

Wheelwright was an English attorney who emigrated to Victoria in 1852, hoping to make his fortune in the goldfields. He had no luck and became a hunter instead, supplying meat and skins to Melbourne. In his memoir, published in 1861 under the pseudonym "the Old Bushman," he described the most effective method of poisoning dingoes, the best time of day to shoot koalas, the proper way to stretch a possum skin, how to bait a platypus (potatoes), how to dig an echidna from its hole, and how to render emu oil. He recalled shooting 168 different species of birds, at least 2,000 kangaroos of all sexes and ages, and, in a single night, 93 possums.

Despite these tallies, Wheelwright seems to have viewed himself as temperate. He complained about the hunters who killed kangaroos "just for the sake of killing them," leaving their bodies to rot. "It will, perhaps, be a matter of regret, at no very distant day, that the kangaroo . . . shall have become, like many animals and birds in the Old World, a theme of bygone days," he wrote. Without laws to restrain them, Australian hunters were acting out the tragedy of the commons: A hunter might shoot an animal during its breeding season even though it was poor long-term policy, he wrote, "when he knows

that they are pretty certain to be shot by some one or other less scru-
pulous than himself."

Lawmakers began to debate what to do. It was easy, one said, "to
utterly destroy native animals, and once that is done they can never be
reproduced." Others were unconcerned. "Of what use is a wombat?"
mused one. "Of what good to the country are a thousand or even a
million kangaroos?"

The colonies' first game laws, passed in the 1860s, aimed to protect nonnative birds that the colonists had introduced into the country. These laws eventually expanded to include native birds, then
native mammals. As memories of the old country faded and a sense of
Australian nationalism swelled, the continent's unique flora and fauna
became a source of pride and cultural identity, Adrian Franklin writes
in *Animal Nation*. This improving attitude, though, did not extend to
the country's venomous snakes.

"Ever since the Devil entered into the Serpent, it became hateful
to all," wrote the English cleric Edward Topsell in his 1608 *History
of Serpents*. This view, Kevin Markwell and Nancy Cushing wrote in
Snake-Bitten, "was carried to Australia, stowed in the cultural baggage
of British colonists, and passed down for generations." In *Venomous
Encounters*, historian Peter Hobbins quoted an Australian settler who
said, "In the bush it has ever been a point of honor . . . to kill every
snake you see, if possible, no matter how difficult the job, nor how
great your impatience to be after other jobs." Another settler suggested in his memoir that the government should offer "some slight
reward, say sixpence or a shilling, for every snake's head that was
brought in." Even people professionally interested in snakes supported the extermination effort. In the introduction to *The Snakes of
Australia*, Gerard Krefft thanked the local men "through whose exer-
tions the dangerous snakes of the neighborhood of Sydney have been
considerably reduced."

Eric Worrell dismissed the old superstitions. "We can't blame the
serpent for having Adam cast out of paradise," he joked. "It was really
Eve's fault." The fear Australians felt about snakebite, played upon by

the wandering snake men peddling proprietary snakebite antidotes, had little quantitative basis, he wrote. "According to statistics snakes kill an average of about five people every year in Australia. In round figures the chances of dying from snake-bite are thus only one in two million."

This statistic was in marked contrast to other parts of the world. Even today, venomous snakes kill more than one hundred thousand people annually. Another four hundred thousand or so are left with amputations or permanent injury. According to a 2019 study, nearly 70 percent of snakebite deaths occur in just five countries: Bangladesh, India, Nepal, Pakistan, and Sri Lanka. The majority of the dead are farmers, and most of them are bitten on the foot or ankle. In 2017, the World Health Organization returned snakebite to its list of "neglected tropical diseases" afflicting mostly impoverished people in the developing world, placing it in the same category as illnesses like dengue fever, river blindness, and leprosy.

Underscoring this disparity is the fact that, as Worrell noted, one of the commonest ways for an Australian to be bitten by a venomous snake is by grabbing it. This remains true. A 2012 study tallied 106 Australian snake handlers who were bitten by venomous snakes between 2004 and 2011. The victims had a median age of forty, and most were bitten on the hand or arm. All but two had been bitten before, and all but two were men. In the United States, where snakebite deaths are equally rare, researchers have offered an even more succinct analysis of snakebite, which they say nearly always co-occurs with a set of personal characteristics that they abbreviate as "the four T's": tattoos, trucks, tequila, testosterone.

"The way to avoid snakebite," Worrell concluded, "is to avoid snakes." In reality, he argued, it was the snakes that needed protection. Australian conservation laws continued to exclude reptiles and amphibians, a loophole that he argued could lead to many species' extinction. "It seems regrettable that legislation is necessary before the Australian public will protect its own fauna," he wrote, "but experience has shown that animals not guarded by law are automatically, though often wrongly, regarded as noxious."

~

After the war, Worrell devoted himself fully to the pursuit of his dream. He wrote about reptiles and hauled barrels of them from the Northern Territory and Queensland south to Sydney. When snakes and crocodiles appeared unbidden in houses and universities, people summoned him. The newspapers covered his every move, calling him "reptile hunter," "snake collector," "professional snake catcher," "an expert in reptiles," "naturalist," "herpetologist," and "snake man." One newspaper reported, "Mr. Worrell claims to have been bitten by snakes on more than a hundred occasions. He treats himself and has never suffered any ill effects from the bites." Another journalist interviewed his mother, who said that Eric was lodging many reptiles at the family home. Aside from the water snake and the baby crocodiles, which shared the tub, these reptiles seem to have been given free range. "It's nothing to find a small snake in his shirt on washing day," Mrs. Worrell said.

In July 1948, Australian newspapers congratulated Worrell and his new bride, Carol Hawkins. Like his parents, she seems to have been accommodating of his dream. Despite her not appearing to have had much prior interest in snakes, the newspapers deemed her a "snake handler" and "snake woman." "Snakes! They don't worry me," she told one reporter. "A lot of people have the wrong idea about snakes," she told another. "They are definitely not slimy. They are silky smooth to touch. I see them as long lengths of beauty." In a photograph accompanying one of the articles, Worrell is wearing a suit and dangling a tiger snake by the tail. His other arm is wrapped around Hawkins, comely, curly haired, cradling a basketful of pythons.

In 1950, Worrell opened his Ocean Beach Aquarium in Woy Woy, an hour by train north of Sydney. As Markwell and Cushing point out in *Snake-Bitten*, this was something of a misnomer, the aquarium being primarily a reptile house. In August of that year, newspapers reported that one of Worrell's taipans was lost in the mail. A few days later, the director of the Sydney General Post Office announced that "snakes or any other reptiles considered dangerous were prohibited as a postal consignment."

In 1951, Worrell entered into a snake-milking contract with the Commonwealth Serum Laboratories, which used the venom he collected to produce antivenom. Australian authorities had begun regularly producing tiger snake antivenom in 1931, and scientists were conducting research that would yield, by the early 1960s, antivenoms for the venom of brown snakes, coastal taipan, and death adders, and, in 1962, a "polyvalent" antivenom, effective against all Australian venomous snakes. Studying and producing antivenom required a steady supply of venom. Historically, this venom was collected by scientists from snakes they either captured themselves or bought from snake men like Snakey George. Worrell, with his wide renown and dedicated facilities, was the first to offer venom by mass production.

In 1952, he published his first book, *Dangerous Snakes of Australia*, an identification aid for taxonomically disinclined readers. That November, while he was away from the Ocean Beach Aquarium at dinner, a drunken park attendee shouted, "I'm not afraid of snakes," then jumped into the snake pit. He picked up a tiger snake and began petting it. A deep-sea fisherman jumped in after him. The fisherman and the drunk waged battle among the snakes until the fisherman finally hoisted the drunk from the pit. "The man will never realize how lucky he was," Worrell told reporters.

Soon after, Worrell made his first TV appearance. "I was all thumbs, missing cues and fumbling words," he later wrote. When he opened the bag holding the snake he intended to milk, he found it was dead. Someone had stepped on it. He milked the snake anyway, jiggling it to give the impression of liveliness. "Miraculously, two drops of venom rolled down the glass beaker," he wrote. "As I bundled the body into the bag, I overheard one of the technicians: 'Must take a colossal nerve to handle a snake like that.'"

After the success of his first appearance, the Australian Broadcasting Corporation contracted Worrell to produce and star in a documentary on catching tiger snakes. Soon he was appearing regularly on TV. One short piece, titled "Operation Crocodile," showed Worrell in Queensland, splashing around in shallow water, trying to catch a crocodile, aided by a one-armed local man. The narrator did not explain whether the man's one-armedness was related to catching crocodiles.

Worrell had a gift for memorable quotes, and reporters relied on him as a public expert on all herpetological matters. He continued to publish numerous works of his own, including the "Don't You Believe It" column in *Australian Country* magazine, dedicated to dispelling myths and misinformation about Australian fauna. "Snakes do not like milk," he reminded readers in one column. "They will not drink it unless water is unavailable." In 1958, he published *Song of the Snake*, a collection of his magazine articles. Reviews were mixed. One said that the book was "not particularly well-written but disarming in its love of Nature and full of fascinating information." Another reviewer wrote that "snake-charmer" Worrell "writes well," but suggested that "most would agree that snakes as a subject for a popular book would be wide of the mark."

The following year, Worrell opened the Australian Reptile Park in Wyoming, ten miles north of the Ocean Beach Aquarium. Markwell and Cushing write that he had originally envisioned this much larger facility as focusing almost entirely on zoological and toxicological research, with tourism a secondary concern. In corporeal form, though, it most closely resembled a theme park, with wandering wallabies and kangaroos, two rows of glass-fronted reptile enclosures, a large crocodile pond, and snake pits where Worrell often entertained guests by milking snakes—a kind of mirror image of the old snake man's act of purposefully taking bites. In 1963, as further enticement, he commissioned an artist to build an eighty-seven-foot diplodocus next to the park's entrance. "It was a masterpiece of engineering, design and sculpture, and its form was graceful and realistic," Markwell and Cushing generously write. Rendered in concrete, the diplodocus was of the early school of dinosaur portrayal, tail-dragging and osteoporotic. It was one of Australia's first Big Things, beating out the famous Big Banana of Coffs Harbour by some months, and perhaps preceding even the Big Scotsman of Adelaide. People came from miles around. Worrell's lifelong dream, which still looked like a dream, was nearly in hand.

~

A few years earlier, Worrell had described his first new species. Looking back, this was an inflection point: the first taxonomic act by an Australian snake man. It was a pioneering, inspirational event, the beginning of a long series of such acts by other Australian snake men, which would culminate in an international debate concerning alleged acts of taxonomic vandalism by Raymond Hoser.

On its own, though, Worrell's act—describing a small snake from Queensland—did not seem very consequential. Relying on just two individual animals, he offered a description of the new species' tooth and scale anatomy, and its color ("dark slatey brown"), and noted that, during the brief period he managed to keep one of the snakes alive, it "displayed no aggressive tendencies and proved to be entirely nocturnal." The new species was easily distinguished from the similar *Glyphodon tristis*, he wrote, by the number of its teeth and rows of scales. He called it *Glyphodon dunmalli*, after Bill Dunmall, who had collected one of the snakes.

Kevin Markwell says that Worrell's foray into taxonomy seems to have stemmed from his interest in snake venom, and particularly in the bite victim's sudden question: What just bit me? "He felt that it was important that venomous-snake taxonomy was well understood, so that if someone is bitten by a particular species of snake, the appropriate antivenom to that particular group of snakes would be administered," Markwell says. "In order to do that, you had to sort out the snake taxonomy. I think it was important to him."

In Worrell's letters and articles, he displayed an easy familiarity with the arcane details of scale, dental, and skeletal anatomy that taxonomists often use in describing snakes and other reptiles. He traveled widely across the country and had more experience with its reptiles in their natural environments than most, if not all, of Australia's museum- and university-based scientists.

But in the decades since he left grade school to begin a string of odd jobs, the science of taxonomy had changed. Where the preceding generations of taxonomists had mostly been amateurs or scientists trained in adjacent fields, after World War II, the science grew increasingly professionalized. Standards of evidence and descriptions of species were far more rigorous and detailed than they had been in the days of Blandowski and Krefft, let alone Linnaeus and Banks. In trying to meet this standard, Worrell repeatedly stumbled.

His first species description suffered from imprecision, Markwell wrote, with a reliance on qualitative descriptions like "moderately long" for anatomical features instead of quantitative measurements. He had bad luck in several subsequent taxonomic efforts, twice describing species only to find that someone else had described and named them a couple of years before. "These are understandable errors for someone who was not a professional zoologist attached to a museum or university and thus without easy access to the technical literature," Markwell wrote. "But they nevertheless worked against his attempts to build his credibility."

Worrell was undeterred. In 1963, he published *Reptiles of Australia*, the first book to tackle the taxonomy of all the continent's reptiles. Despite the breadth of its subject matter, he was clearly most interested in snakes, devoting much of the book to their natural and cultural history and to dispelling myths. "A resident of the Gosford district tells me he saw a hoop-snake at Milperra, New South Wales," he wrote. "The snake, brownish, about three feet long, was bowling rapidly along, tail in mouth—in the form of a circle—in pursuit of a rabbit. The circle was inclined to sag a little. After rolling about fifty feet the snake suddenly released its tail and sprang forty feet to strike the rabbit, which gave a squeal and then died. This man, whose father and brother, he said, also saw it happening admits that the occurrence took place many years ago. Over such a long period memory can be unreliable."

The heart of the book was a taxonomic checklist of Australia's reptiles. This was Worrell's assessment of all the hypotheses about the various species and relationships among them that taxonomists had offered up to that point. Some of these hypotheses he dismissed, ignoring species and genera proposed by other authors, including by Roy Kinghorn, the curator of birds and reptiles at the Australian Museum and a longtime correspondent. Others he elevated, resurrecting species that authors of earlier works had dismissed. In total, Worrell tallied 107 species of snakes belonging to 55 genera, including 3 species and 9 genera he had described himself.

Professional taxonomists reviewed the effort poorly. "The purpose of *Reptiles of Australia* and the audience for whom it was intended are unclear," one herpetologist wrote. "I believe the book has fallen far short of its purpose. The numerous errors and inconsistencies in presentation will preclude it from being a valuable reference for the layman, casual student, field naturalist, and professional alike." He noted omissions, bad photography, and the troubling absence of a bibliography. Another herpetologist wrote that "Worrell's book is clearly the work of an amateur, being strewn with mistakes and inaccuracies, and best left ignored by serious herpetologists."

Worrell published just two more taxonomic works, one describing a new snake genus, another describing two new snake subspecies. Both descriptions appeared in *Australian Reptile Park Records*, a journal that Worrell published, and whose two and only issues consisted entirely of those descriptions. While these documents technically satisfied the requirements of the International Code of Zoological Nomenclature at the time, it was an unusual route to publication, circumventing the typical process of independent peer review. Publishing descriptions in his own journal inevitably looked like self-dealing, and it undermined any desire he had to be taken seriously by professional taxonomists.

When university- and museum-based taxonomists rejected the

taxonomic efforts of later snake men, it would spur them to action, rebellion, even retribution. But Worrell seemed to take a measured view of the situation. In his 1966 book *Australian Wildlife*, he poked fun at himself, a jack-of-all-trades in a world of specialists. "I enjoy working on a book like this—with the help of my many friends. In a strange way they have confidence in what I do," he wrote. "Some are scientists: they say my work is artistic. Some are artists: they say I am a perceptive zoologist. Some are photographers: they admire my literary style. Some are writers: they like my photographs."

The transcript of the meeting of New South Wales's legislative body on August 18, 1970, records the questions of one Mr. Humphries, who was concerned about snakes. Was the premier aware, Mr. Humphries wanted to know, that there were an estimated three hundred collectors and dealers trading in venomous snakes in New South Wales? Did the premier know that snakes imported into Australia from other parts of the world could "adjust to climatic condition here, breed, and become a menace of much greater significance than the rabbit?" Mr. Humphries requested that, "in the interests of public safety," the premier introduce legislation that would regulate the housing, keeping, and transportation of "dangerous" reptiles.

The premier, Mr. Askin, agreed. "I shall consult the appropriate authorities to ascertain the most satisfactory action," he said, and promised to return once he'd gathered "further authentic advice."

A little more than a month later, Mr. Askin reported to legislators that he had come up with some answers. On the question of whether exotic venomous snakes might become a rabbit-like plague, the government had consulted "acknowledged expert" Eric Worrell, who said that it was unlikely. "It appears that snakes are not intelligent creatures and if placed in a strange area cannot establish themselves satisfactorily," Mr. Askin explained. "They refuse to eat and pine away."

Worrell was near the height of his fame. Earlier that year, Queen Elizabeth II had awarded him membership to the Most Excellent

Order of the British Empire for his contributions to herpetology, one of the rare occasions when the royal family has shown interest in snakes and lizards. But while his renown continued to grow through the early 1970s, he was increasingly dogged by troubles.

The first problem, as is often the case with theme parks, was financial. The Australian Reptile Park had always relied on its snake-milking operation as a prime source of revenue. But Worrell was reluctant to raise the prices he charged the Commonwealth Serum Laboratories, and by the late 1960s, he was selling at a loss. Meanwhile, Markwell and Cushing write in *Snake-Bitten*, the costs of feeding his swelling animal collection had grown. The animals required a steady stream of "mice, fowl, pigeons, chicks, eggs, fish, rabbits, guinea pigs, rice, fruit, horsemeat, liver, prawns, milk and cod liver oil," as well as fresh or frozen pythons for the king cobras, which prefer a diet of other snakes. Further trouble was caused by the park's employees, who, in bald defiance of Worrell's vision, voted to unionize. He fired the mutinous workers and then was forced by a labor commission to rehire them; after the incident, turnover at the park increased and remained high for the rest of his time at the helm.

Drinking, which had long been one of Worrell's favorite pastimes, was fast turning into a second job. His drunkenness increased notably after his divorce from Carol Hawkins in 1970. In the late 1950s, her brother had drowned in the Johnstone Gorge while on a trip with Worrell. "His missus blamed him for it, and eventually left him over it," one of Worrell's friends told me. Worrell appeared so regularly at the Grange Hotel across from the reptile park that a corner of the bar was named in his honor.

Another problem came in 1974, in a law that affected how he could obtain new animals for the park. In the early 1970s, when the government of New South Wales began considering protections for the state's reptiles and amphibians, it again consulted Worrell, enlisting him in its herpetological advisory committee. Markwell and Cushing write that he provided a list of rare and dangerous reptiles that he thought only licensed professionals should be able to collect and keep, as well as a list of common and harmless species that he thought should remain available to amateurs. Legislative discussion focused mostly on the bill's other provisions, which included adjustments to national parks, Aboriginal sites of cultural importance, and parking enforcement. Toward the end of a lengthy speech, legislator L. A. Solomons mentioned his satisfaction that the bill included protections for reptiles, although he had mixed feelings about the provision that "carries forward the old biblical injunction against snakes, and provides that it is still not an offense to kill snakes." He continued, "All snakes have this carry-over from our religious background, which has lasted since the days of Eve."

"That was not a snake," C. Colborne interjected.

"It was a sort of snake," Solomons replied.

In January 1974, *The Sydney Morning Herald* announced new regulations set to go into effect that June. With the exception of nine especially common species, it was now a crime to collect, kill, or keep reptiles in New South Wales without a license. The aim, National Parks and Wildlife Service director D. A. Johnstone told the paper, was "to stop amateurs, particularly children, from owning reptiles and neglecting them through ignorance." He added that the service would not prosecute anyone for "killing or attempting to kill a snake."

~

When Kevin Markwell arrived home from school, his mother handed him the newspaper article announcing the new regulations. "I couldn't believe it," he later recalled. "Keeping reptiles was who I was. I took the article and headed into my room, closed the door, and began to sob." He was thirteen. Many of the reptiles he kept, including his three bearded dragons—Cuthbert, Horace, and Cordelia—were now illegal to keep without a license. Reptile kids, already quirky and unusual, were pushed to the legal fringes, Markwell wrote, "subjected to significant fines and public humiliation."

This is how countless stories and legends begin: An ordinary, righteous person finds themself suddenly in violation of the law, driven from their former path and forced into action. Only then, from the outside, can they see the systems and institutions that surround them for what they really are—faceless, indifferent, iniquitous. In Howard Pyle's retelling of the most famous version of this tale, a young Robin Hood is walking to Nottingham to enter the sheriff's archery contest when a band of drunken foresters goads him into killing one of the king's deer. Realizing his error, Robin flees. One of the foresters shoots an arrow at him and narrowly misses. He returns an arrow and does not miss. "And so he came to dwell in the greenwood that was to be his home for many a year to come," Pyle wrote. "For he was outlawed."

Unlike Markwell, Worrell seems to have not immediately realized the change in his position. To the contrary, the passage of the law must have seemed like a victory for the avowed conservationist. He had spent decades advocating for reptiles and amphibians and other "unpopular ones," as he described them, arguing that they should be given legal protections. In time, though, the new law exposed a central contradiction of Worrell's livelihood: Despite his publicly stated views, his business was a furnace fueled by fresh animals.

An Australian Reptile Park inventory from the 1970s included a frilled lizard, monitor lizards, legless lizards, blue-tongues, shinglebacks, a gecko, a tuatara, water dragons, Gila monsters, iguanas, skinks, bullfrogs, alligators, crocodiles, boa constrictors, anacondas, pythons, rattlesnakes, swamp snakes, mangrove snakes, tiger snakes, black snakes, brown snakes, a sidewinder, death adders, puff adders, a rhinoceros viper, a Gaboon viper, pit vipers, bushmasters, a black mamba, taipans, king cobras, kraits, kangaroos, koalas, wallabies, wombats, and platypuses. The park suffered constant attrition. One of the more notable losses was of Casanova, a fifteen-foot freshwater crocodile, which died just three weeks after Worrell brought it to the park. He told reporters that vandals had entered the park and shot Casanova between the eyes with a .22 rifle, although Markwell and Cushing note that at least two park employees had reported seeing blood coming from the crocodile's mouth even before the bullet wound appeared. They suggest that the animal died from the stress caused by its captive journey from Australia's north, and point out the mysterious disappearance of its carcass. "Perhaps," they write, "Worrell sold Casanova's body to someone looking to own a large mounted crocodile."

Collecting snake venom also took a heavy toll on the animals. The milking process was stressful, often fatally so, and acquiring the thousands of milligrams the serum labs needed to produce antivenoms required a constant supply of new snakes. "I think he became desensitized to the deaths of reptiles," Markwell says, "and he kind of believed his own rhetoric, that these reptiles had to be sacrificed in order to supply venom for antivenom purposes."

For a while, Worrell's public stature led authorities to accommodate him, and he continued to collect snakes and other animals from the wild. By the early 1980s, though, the New South Wales National Parks and Wildlife Service began requiring him to go through the same licensing process as other members of the public. "He was really pissed off that he no longer had an open carte blanche to collect as many snakes as he wanted to, whenever he wanted to," Markwell says. "He'd had that ability for many years, and then those doors started to shut in his face."

Worrell wrote a letter to his member of Parliament, then to the prime minister, arguing that the reptile park should be viewed in relation to the lives it had saved. In a video from this time, he sits on a park bench, kangaroos meandering behind him, his head encased in a shimmering white halo of his own hair. "It has been estimated that we save one life a day from snakebite," he said. "I think that works out to something around twenty thousand lives that this organization has saved since we started." Given the rarity of snakebite in Australia, this claim can be taken less as a statement of objective fact and more as an impressionistic plea for his park's value to society, a reflection of Worrell's view of things: He was a hero, but the authorities were treating him as a bandit.

His problems were mounting. As he struggled to acquire venomous snakes, the once-brimming snake pits emptied. Fewer visitors came, and food and gift-shop sales declined. Snake-milking contracts dried up. The cost of "hay, grains, fruit, mice, mealworms, crickets, locusts, earthworms, silkworms and day-old chicks" increased by nearly half in just one three-year stretch, Markwell and Cushing write. The park's debt grew, with a corresponding increase in interest payments. Repairs and upkeep went undone, "leaving the park run-down and unkempt." Worrell tried again to lobby the government for a property-tax waiver, but was denied. He raged against federal paperwork and fought the local council. The New South Wales National Parks and Wildlife Service audited the park, finding it held nearly two hundred animals that weren't in its records, noting that the fatality rate of its snake-milking operation was unusually high and that the park did not possess a necessary fauna-dealer's license. "It was a time of great crisis," Markwell and Cushing write.

During this period, Worrell was arrested twice for drunk driving. He was hospitalized repeatedly but continued to deny having a problem. As he spiraled deeper, his second wife, an assistant at the park, left him for a reptile keeper. One of his friends told me he discovered him one night outside the park, raging drunk, waving a pistol and yelling about the reptile keeper. His friend wrestled him across the street and installed him at his usual seat at the bar. In 1985, a cobra bit Worrell on the thumb while he was milking it in front of 150 people, nearly killing him. The next year, he relinquished ownership of the park to his second wife and the reptile keeper. The year after that, he had a heart attack and died. He was sixty-two.

One day late in the southern summer, I traveled an hour north of Sydney to Somersby, current site of the Australian Reptile Park. Worrell's second ex-wife and the reptile keeper moved it there in 1996. This iteration of the park is orderly, clean, distinguished by its extensive use of sculpted concrete. There are concrete igneous, sedimentary, and metamorphic rocks; concrete biological forms; concrete culturally derivative motifs; and, lurking over the park entrance, a concrete frilled lizard. I spent the morning wandering around the exhibits, watching keepers give reptile demonstrations that were educational but without suspense and that culminated in a twenty-nine-dollar photo opportunity. I bought some fries at the Hard Croc Café, looked at the Komodo dragons, and then went and gazed for a long time over the alligator pond, feeling lonesome, considering a swim. I continued to the reptile house, which was dominated by an enormous anthropomorphic concrete crocodile in pharaonic regalia that every minute or two offered deafening lessons in the sonorous timbre of James Earl Jones. "Within the walls of the lost kingdom of reptiles," the crocodile said, "you will observe grace, beauty, power, danger, specialization, and *adaptability*." After passing through the gift shop, stocked mostly with books by and about the reptile keeper, I left.

I walked a mile down the road and climbed a steep rocky path off the Pacific Highway. At the top, after aspirating a large, glutinous spiderweb, I found Ploddy the concrete diplodocus. Once an irresistible magnet to Australian reptile kids, Ploddy is now painted bright yellow, surrounded by a barbwire-topped chainlink fence, and dimly visible to cars passing on the highway below. This, in physical form, is all that remains of Worrell's dream.

Eric Worrell achieved remarkable success, built on the back of a revolutionary message: that, contrary to thousands of years of mammalian instinct and cultural enmity, snakes and reptiles were valuable, admirable, even lovable. Possessing an unusual energy and determination, he spread this message in articles, books, TV appearances, and documentaries, all the while building his park and working as a major venom supplier. By his charisma, he dragged every other snake man up out of the dust and toward the light; today, the most workaday snake man is conversant in the basic biology and ecology and taxonomy of his charges, in the importance of his craft to conservation and public safety. While Worrell is mostly forgotten outside the narrow world of Australian herpetology, eclipsed by the snake men and crocodile hunters and lizard people who came after, he remains everywhere just below the surface, so central that the story doesn't make sense without him.

But as is often the case with transformational, Promethean figures, his story is ultimately a tragic one. As Markwell sifted through Worrell's documents and talked to the people who knew him, he came to see what was hidden by Worrell's heroic veneer—not only alcoholism, but hubris, arrogance, greed. These lie within the typical suite of human flaws, but to me, Worrell's story also seems to hint at something deeper, at the danger that can lie in the single-minded pursuit of a dream. More than once, he seemed almost to grasp it, only to find it could not be held. It grew and grew, consuming animals, money, and relationships, swelling without definition or limit, until it pressed against the bounds of practicality, ethics, the law.

Despite his frustration and anger at the injustice of the institutions that constrained him, Worrell mostly followed the rules, losing himself in an alcoholic haze. His only real rebellion was against himself. Others who followed were less acquiescent, choosing a life in the greenwood rather than bowing to the authorities. Two of them are men Worrell knew, men who became renowned and reviled for their rebellious taxonomic acts, although they have never stopped insisting upon the righteousness of their cause. Like Bonnie and Clyde, Butch Cassidy and the Sundance Kid, or Robin Hood and Little John, these men are recorded in history most often as a pair: Wells and Wellington.

CHAPTER 6

WELLS AND WELLINGTON

"The reason for that," said Sansón, "is that printed works are read at leisure and their defects are easily spotted, and the more famous the author the more closely they're scrutinized. Men renowned for their genius—great poets, illustrious historians—are usually envied by those whose pleasure and pastime is to pass judgement on what others have written, without ever having published anything themselves."

"That is not surprising," said Don Quixote, "because there are many theologians who cannot preach, yet are experts at identifying the faults and the excesses of those who can."

—CERVANTES, *DON QUIXOTE*

One night, years before the Wells and Wellington Affair, Richard Wells was driving in his car along New South Wales's North Coast, looking for snakes. With him was Grant Husband, who he says is a nice man and who plays no further part in the story, and Peter Rankin, who does: Both Wells and Wellington say that, if it weren't for Rankin's tragic and untimely death, things would not have gone so sideways between them and the Australian Museum, that conflict being what forced Wells and Wellington to strike out on their own and publish their famous and controversial taxonomic treatises.

On the night in question, Wells, Husband, and Rankin were driving slowly through the coastal rainforest, stopping along the way to investigate any crawling or creeping thing that appeared in their headlights. They were close to a small town where they planned to get dinner. At around 7:30 p.m., Wells spotted a light behind them in the distance, which he thought was another car. It was unusual to encounter other drivers in that part of the country but not impossible. As Wells watched, though, the light disappeared. He thought the car had gone off a cliff. Then the light reappeared. The three men pulled to the side of the road to let it pass. When they stopped, the light stopped. They started and the light started. They stopped again and so did it. It disappeared and reappeared and disappeared again. Wells got out of the car and walked back down the road to where he'd last seen the light. In the beam from his flashlight, he could see only one set of tire tracks. He walked back up the road. There must be another road somewhere we don't know about, he said to Rankin and Husband. They continued toward the town.

Soon the light reappeared, this time above them to their right, in the treetops. Wells, at that time still bound to a painfully literal view of reality, suggested again that there was another road somewhere parallel to the road they were on. Or maybe it was a helicopter. They stopped the car to listen. There was no vehicle sound, no thrum of rotors. Then the light emerged from the forest and swept across the coastal plain, then over the sea, and then it vanished. A small clock was superglued to the dashboard, Wells recalled. "We could see the time, and it was quarter to eight," he said. "And we were in the process of talking to one another and all of a sudden . . . "

Something happened. He did not actually tell me what happened, because he says it was beyond comprehension, and would be especially incomprehensible to someone with generally unremarkable powers of comprehension. "But I'm telling you now," Wells told me, "something happened." Grant Husband, via email, said he remembers the mysterious light distinctly, and affirmed that something happened, but could offer no further information about exactly what happened, since soon after the light appeared, he fell into an impenetrable slumber. Whatever the details of the mysterious event, which lasted until four in the morning, it changed Rankin and Wells. In its aftermath,

Rankin was suddenly full of delightful and uncanny ideas about his herpetological research. He could quickly grasp the most arcane and complicated concepts. He had been smart before, but now he was supersmart. "He said, 'I don't know how the hell I know this,'" Wells says. "But suddenly he knew."

Wells, meanwhile, was imbued with new powers of concentration and understanding and multitasking and found that he could read in foreign tongues. Strangest and most wonderful of all, he gained the gift of foresight. Sometimes it extended only a moment or two, telling him to brake just in time to avoid an oncoming car. Other times the foresight went further in time, as it did with his premonition of Rankin's death, and as it does now, with his foreknowledge of the world's impending renewal.

He also had some idea, of course, long before he and Wellington put names to hundreds of new species and wholly reorganized the taxonomy and nomenclature of Australian herpetology, that their decisive and well-intended act would upset the academics and powers that be. But even he was surprised by what followed.

~

Wells became a snake boy in the late 1950s, when he was eight years old. He'd discovered a snakeskin in the garden where his family lived, in Bundeena, New South Wales, and brought it inside to show his mother. Being possessed of nearly perfect memory, he says he can still clearly picture the skin and now knows, from the shape of the scales and other details, that it belonged to a red-bellied black snake, *Pseudechis porphyriacus*. A fellow boarder in the house was unimpressed. The boarder said the snake it belonged to was dangerous and that bringing its skin inside would encourage the snake to follow. He whipped Wells with his belt. "Bloody hell, this herpetology's got a bad start," Wells recalls thinking. He ran outside and went looking for the snake itself.

His father was a sheep shearer, then a scrap metal dealer, then a machinery merchant. The family moved frequently, often landing in houses on the outskirts of town, surrounded by bush. "I was constantly exploring wherever I was," Wells says. One of his favorite activities was to walk the banks of creeks back to their source, to discover the land's hidden connections. "I was naturally interested in nature," he says. "It was, I think, a distraction from the extreme poverty that I was in at home."

In 1962, the family visited Eric Worrell's Australian Reptile Park. While Wells basked in the park's herpetological riches, his mother struck up a conversation with Worrell's mother, Rita, who was often at the park. The two women became friends, and the family began to make regular visits. During these trips, Worrell, the famous snake man, always found time to talk to Wells, the snake boy. "I always found him very personable and very readily willing to share information, at least to a kid," Wells told me.

By the time he was a teenager, he'd attended a dozen different schools. He left school for good at fourteen and got a job at a garbage dump—"rubbish tip," in Australia—sorting scrap metal, three dollars a day, sunrise to sunset. Buried there among the decrepit furniture and broken bottles and every conceivable type of junk were many books. He acquired a large English pram with big wheels, and every day he filled it with books. "People would watch me going past with a pram, and they didn't know what I was doing," Wells says. "And some people thought, 'What the hell is this bloke doing with this pram walking up and down this hill?'" By the end of the year, he had a thousand books.

Around the same time, he began collecting specimens. At first, they were animals he found dead alongside the road. He read Pulvertaft and Judah and Kaiserling and other great physicians and anatomists and preservationists, and he learned the arcane art of pickling flesh and developed his own special formulas. Other people heard about his collection and started saving him the dead animals they found. "I'd go around to their places and pick up their jars of dead reptiles," he says. "Bring out your dead!" He studied the animals "to see what they'd been eating, how many babies they had in them, what parasites they might have had, all ecological stuff." The collection grew and grew.

In the years that followed, Wells worked for a telephone company, for a hospital, for scientific laboratories. He aimed to work a hundred jobs by the time he was thirty, his way of seeing the world. Any job would do, so long as it left him time to make field trips. He traveled across Australia, visiting what seemed like every little dot on the map, studying nature, examining its creatures, always learning. He collected more books and more specimens. Eric Worrell began to give him the bodies of animals that had died at the Australian Reptile Park, and now when they talked to each other, it was not adult to child, but snake man to snake man.

In 1974, New South Wales passed the National Parks and Wildlife Act. This was the same law that would eventually restrict Eric Worrell's ability to collect reptiles from the wild and turn reptile kids like Kevin Markwell into outlaws. After the law passed, but before it was enacted, Wells asked the permitting agency whether he would be able to get licenses for the preserved reptiles in his collection. Many of them were suddenly illegal to own, and Wells faced potentially hundreds of thousands of dollars in fines. "I asked them, 'Can I get a permit to have these dead snakes and lizards and so on?'" Wells recalls. "They said, 'Oh yeah, sure.'" But the permit forms never arrived. Months passed. On the day before the law was set to take effect, he called the head of the licensing section. The license head apologized but said Wells's only option to avoid the fine was to surrender his specimens to the Australian Museum.

Wells and his father loaded the collection, which he says contained ten thousand specimens, into a truck. "I had them in immaculate preserving bottles," he says. "It cost me huge money for all those beautiful jars. Everything was magnificently labeled. I took them all to the museum and handed them over. My beautiful collection of preserved reptiles went to the museum." He was angry, but his anger was tempered, first, by the fact that it had always been his intention to eventually donate his collection to the museum, and second, by the museum staff's assurances that he would still be able to study the specimens. "I was told that I could have access to my collection," he says. "I was naïve, really naïve."

~

Ross Wellington grew up in the Sydney suburbs. He caught his first frog in the late 1950s, when he was around four years old. The frog was in a cistern at a Chinese market near his house. It was a green and golden bell frog, at that time one of the region's commonest frogs. As he grew older, he kept cicadas and silkworms, jacky lizards and bearded dragons and blue-tongues, parakeets and lorikeets and other types of parrots. He climbed trees to steal eggs from the nests of wild birds, which he notes was perfectly legal back then. He visited the Australian Reptile Park and watched Eric Worrell milk venomous snakes. He liked snakes, but his interests were broader. He wanted to be a biologist.

For years, the dream didn't change. Just before he was set to enter university, he got his girlfriend pregnant. They married and moved into his wife's parents' house in Wollongong, an hour and a half south of Sydney. He took a job at a steel mill and entered Macquarie University. It was a difficult time, he says, commuting and working, adjusting to marriage and fatherhood. His university classes were a respite. "For me, studying biology was almost as, I imagine, some people's drug trip," he says. "It was euphoric. I just loved it."

In physics class, he met Peter Rankin. They both hated physics and loved biology, Wellington says. Rankin was focused on reptiles and displayed an unusual talent for taxonomy. Wellington's main interest was in the emerging field of genetics. The two would talk for hours and go on field trips together. "I can remember catching twenty-six species of frog in one night," Wellington says. "Rankin was my best mate."

Rankin took Wellington to the Australian Museum, where he'd been working in the herpetology department, and to meetings of the Australian Herpetological Society, a mostly amateur organization founded in 1949. At that time, Richard Wells was society president. Wellington recalls that Wells "was like Peter—absolutely fanatical herpetologist—but like Peter on steroids. He takes everything to a degree that's beyond most people's imagination." After one of the meetings, the three of them went out to dinner together. The conversation, Wellington says, was "like petrol and matches."

Later, Wells took Wellington to see his library, which even then was enormous. It included "every single paper document that had ever been written about Australian reptiles," Wellington says. "He'd fanatically gone, like, you know, white ants eating wood, finding where these obscure things were hidden away." Wellington helped Wells organize the library, "trying to get it into a rational state."

Near the end of his time at university, Wellington took a temporary job in the ornithology department at the Australian Museum, entering written records into the computer, roughly equivalent to Rankin's job in the herpetology department. Wells was also at the museum, studying the specimens he'd forfeited after the 1974 law came into effect. He was "like a piece of the furniture," Wellington said. "He was always there, counting scales and measuring stuff. He'd be there at 8:30 in the morning when you'd get there, and when you go at 5:30 in the evening he's still there." The future seemed assured, Wellington says. Rankin would keep working at the museum, becoming a leading Australian herpetological taxonomist, perhaps rising to curator. Wellington, meanwhile, would become an expert in the burgeoning field of genetics. Wells, too, was clearly bound for greatness, of the unusual, paradigm-breaking variety.

Then, in the late 1970s, Wells organized an Australian Herpetological Society field expedition to New Caledonia, a small archipelago 750 miles east of Australia. "It was like Pandora's box," Wellington said of the archipelago's reptiles. "It was a unique assemblage of reptiles that no one had done virtually any study of." For reasons that remain unclear, Rankin did not wait to join the group trip, set for January, but instead organized his own trip for December, with another young herpetologist named Ross Sadlier. Wells says he could predict what would happen. "I said, 'Look, I can see what's coming. Don't go, Peter,'" he recalls. "And he said, 'I know, but I've got to go.'"

Wellington, who had no premonitions, was about to leave with his wife on a belated honeymoon to Indonesia. "The last words between Peter and I were, you know, 'Good hunting,'" Wellington says. "And he said 'Good hunting' to me." A month later, when he returned from Indonesia, he called Rankin. The person who picked up the phone told him Rankin was dead. On New Year's Day, 1979, Rankin climbed a banyan tree while looking for lizards and fell, breaking his neck. He was twenty-three.

Wells and Wellington both mark Rankin's death as a turning point. It "changed the whole course of Australian herpetology," Wells said. Wellington agreed: Rankin would have smoothed things over, preventing the disastrous schism with the Australian Museum. "If he hadn't died," Wellington said, "Wells and Wellington wouldn't have happened."

The curator of herpetology at the Australian Museum at that time was Hal Cogger. Within the world of Australian herpetology, he was among the preeminent men of science. As a child, he was interested in animals, especially reptiles, an interest that remained with him into high school, when he became an apprentice preparator at the Australian Museum, stuffing animals, taking photographs, and carrying out other odd tasks. Within a decade, he had risen to become curator of the department of birds and reptiles. In the 1960s, when Roy Kinghorn, a former department curator, revised his book *The Snakes of Australia*, Cogger assisted him with the latest nomenclature. In 1974, Cogger published his own book, *Reptiles and Amphibians of Australia*.

It was a broader work than Eric Worrell's *Reptiles of Australia* from a decade before, although Richard Wells says that, as a work intended for a general audience, it was less fun to read than Worrell's text, and that it, too, contained mistakes and omissions. Still, he says, everyone knew that Cogger's book was only a prelude to a much bigger, more comprehensive work, what promised to be the authoritative tome of authoritative tomes, one that would mend all holes in the taxonomy and nomenclature of Australian reptiles and amphibians. "It was all going to be corrected," Wells says. "Everyone was waiting for this."

Specifically, Cogger's forthcoming work would contain the results of his yearslong hunt for synonyms—different names given to the same life-form. As Cogger later wrote, scientists had given the perhaps eight hundred or so actual described species of Australian reptiles and amphibians more than two thousand different names. *Acanthophis*, the death adder genus, for instance, had been given four other names. Meanwhile the species *Acanthophis antarcticus* had eight other names. Figuring out which purported species were real and which were not meant digging through centuries-old books and papers, trying to determine which names were published first. It also meant studying type specimens. This is another requirement of describing a species: It must be tied to an individual creature. Other taxonomists use this creature the way judges in dog competitions look to the breed standard, as a kind of ruler by which other individuals may be measured to determine whether or not those other individuals belong to the same species as the type.

Most of the type specimens of Australian animals were sitting in European museums and universities. Over the centuries, many had been lost or damaged; many others were attached to descriptions that, Cogger wrote, were "brief to the point of obscurity." He and his wife Heather spent months traveling across Europe trying to fill in the holes, examining pickled reptiles and amphibians, weighing whether they were what their description said they were. Cogger told me his hands were "constantly wet with alcohol," so Heather took notes.

The next part of the story is disputed. Wellington left the museum first, when the temporary job was completed. He applied for a permanent position in the herpetology department—a position that he said would have certainly gone to Peter Rankin if he was alive—but the job went instead to Ross Sadlier, Rankin's companion on the ill-fated trip to New Caledonia. There are slightly differing versions of why Wells left. Wellington says it was because the museum cut off Wells's access to his specimens.

Given that these specimens are what the two men say precipitated the whole Wells and Wellington Affair, what follows is a puzzling discrepancy: Wells told me that on the day in 1974 when he turned over his ten thousand specimens to the Australian Museum, he found himself in an elevator with Hal Cogger. "He said, 'This is appalling,'" Wells recalls. "He said, 'Look, Richard, I can't believe what they've done to you. They're absolutely fucking bastards.' I think it's the only time I ever heard Cogger swear, was that time in the lift."

Cogger, who in my interactions with him did not seem like much of a swearer, could not recall this conversation. "It's the first time I've heard of it," he told me. He also did not remember Wells's specimens. "I have no recollection at all of a large bulk of specimens being given to us," he said, "and certainly not under those circumstances." Australian Museum records show that the herpetological collections contain nearly four thousand specimens collected by Wells, the first of which was collected on August 26, 1967, the last on February 21, 1984. The records show only the date of collection, not the date of their donation to the museum, but, if these records are to be believed, only seven hundred of the specimens were collected before 1974. (Wells says that his figure is correct but that since his collection contained specimens collected by other people, they may not have shown up under my search criteria; he says some of the roadkill and other less perfectly preserved specimens that he gave to the museum, meanwhile, were given away or cut up by ecologists. The Australian Museum's media team and herpetologists did not respond to my repeated requests for clarification.)

Wells and Wellington say it was Ross Sadlier who choked off Wells's access to his specimens, increasingly turning him away at the gate. If Peter Rankin hadn't died, they say, it would have been him, not Sadlier, controlling access, and he would have welcomed Wells. An adversarial relationship would never have developed. Their taxonomic work would have come from inside the Australian Museum, not outside. They would have been congratulated for their contributions to Linnaeus's great tower, not accused of trying to tear it down. Sadlier, after responding to my initial email, fell silent when I followed up, and did not offer his own version of events or refute Wells and Wellington's.

Wells says that in the months leading up to his departure from the museum, he was suffering both from the effects of a difficult breakup and also, possibly, from chronic formalin exposure, which he says may have contributed to his depression, making him more volatile than usual.

Cogger says that the reason Wells left the museum was that he "ended up having a dispute with one of our staff members, which ended unhappily, with him departing and slamming my door behind him. And that was that."

This last part is how Wells remembers it, too. "I slammed the door behind me so hard that it came out of the wall," he says. "I don't know, it was an old door."

A little more than a year later, Cogger published his tome, titled *Zoological Catalogue of Australia, Volume 1: Amphibia and Reptilia*, coauthored with Heather Cogger and his research assistant, Elizabeth E. Cameron. With 313 pages of densely set type, the volume contained every described genus and species of Australian amphibian and reptile according to Cogger's best estimation. It also listed all of the other names, or synonyms, that scientists had given those creatures, along with the books or papers that held the synonym descriptions, and the location and collection numbers of their type specimens. These bibliographical details would be of little interest now except that Cogger says they helped set in motion what happened next.

When Wells read Cogger's *Catalogue*, he was shocked. "I saw it and said to Ross, 'My God, he hasn't fixed it. He hasn't done it,'" he says. Everywhere the pair looked were gaps, holes, species unnamed, genuine species relegated to synonymy. There was no more time to wait. Wells and Wellington would have to fix the problems in Australian herpetological taxonomy themselves.

In his book *A Criminal as Hero*, about Angelo Duca, a Robin Hoodesque figure from eighteenth-century Italy, Paul F. Angiolillo writes that the so-called bandit-hero "seems to occur when living conditions among the larger mass of a society are such that frustration, anger, fear, insecurity, poverty, discrimination, protest, and lack of hope are widespread among the people. Political corruption or ineffectualness, economic hardship, and discriminatory social justice most often create or accompany such conditions. It is then that the individual renegade . . . takes to the hills or highways with a band of like-spirited companions."

By the early 1980s, a similarly dark outlook was settling in among the people at work on Adam's task. They confronted a fundamental shift. In the days of Linnaeus, taxonomy had been a basic science—that is, aside from a vague biblical directive, there was no particular practical need to identify, describe, name, and sort out the relationships of all of the world's species. A taxonomist might justify their work in roughly the same way that NASA administrators justify their work to Congress or Mallory justified his lust for Everest—some things, like the workings of the far distant universe, like the highest point on Earth, are simply worth knowing.

While many taxonomists of the 1980s continued to insist that theirs was a basic science—indeed, as many taxonomists continue to insist today—it was increasingly obvious that it had urgent practical applications. The information it provided was of enormous and even existential consequence. Humans, scientists thought, were driving countless species to extinction, forever altering the evolutionary path of life on Earth. Conservation laws and treaties that aimed to address this disaster—including the U.S. Endangered Species Act of 1973, the Convention on International Trade in Endangered Species of Wild Fauna and Flora, and New South Wales's National Parks and Wildlife Act—depended on the ability of enforcers to identify which individual creatures were members of a protected species and which weren't. For a species, remaining undescribed could mean going without legal protection, which could lead to overharvesting or the destruction of its habitat. Going nameless could mean extinction. Taxonomy was once a gentleman's pastime, a matter of curiosity. Now it was a race against time.

Wells and Wellington's exile from the Australian Museum, by this light, was not merely the dismissal of two young employees from entry-level jobs. The stakes were infinitely higher, concerning the very existence of the potentially hundreds of undescribed species among the thousands of specimens they no longer had access to—species that other scientists might overlook for years, decades, even centuries, during which time those species might go extinct.

In *A Criminal as Hero*, Angiolillo writes that one of the key characteristics of the bandit-hero is their role as "great protector," shielding allied peasants, serfs, and peons from the depredations of lords, sheriffs, tax collectors, and other institutional figures. So it was with Wells and Wellington, who say that what they did next was only ever in service of conservation, of describing, naming, and protecting species that had no other defenders. If what they did was banditry, as many of their critics said it was, then it was banditry in service of a higher cause.

~

The pair holed up in Springwood, New South Wales, near where Wellington was working as a high school science teacher. Filled with righteous anger, certain of their cause, they set to work. Like Joseph Smith dictating the contents of the golden plates, Wells recited all the taxonomic knowledge he had accumulated over years of ceaseless study. Wellington recorded these herpetological revelations on printer sheets, on tissues, on toilet paper, on whatever was at hand. They connected these materials sometimes with staples, sometimes with glue. When they perceived a gap in what they had already done, they cut the document in two and spliced the new material in. They worked for twenty hours, forty hours, eighty hours without rest, until Wellington fell unconscious on the floor. "I just looked at him and thought, 'You are pathetic,'" Wells recalls. When the work was complete, he says, they rolled it up, "like a Dead Sea scroll." What had taken Cogger a decade, Wells and Wellington remade in a matter of days.

At first, the two men considered publishing their descriptions in the usual manner, submitting a series of individual papers to scientific journals. But that would have been difficult, Wellington said, given the number of species involved, the sense of urgency they felt, and their status as relative unknowns. There was another avenue. Wells was at that time editor in chief of the *Australian Journal of Herpetology*, a scientific journal he founded during his brief enrollment at the University of New England in north-central New South Wales, during which time he also converted his dorm room to a tropical jungle and accrued more than $800 in parking fines. The first two issues of the journal contained articles by other authors, including Garry A. Webb's "Observations on the Diving Ability of a Red-Bellied Black Snake" and Karen Kool's "Is the Musk of the Long-Necked Turtle, *Chelodina longicollis*, a Deterrent to Predators?" The third issue, published on New Year's Day, 1984, consisted entirely of Wells and Wellington's new work, which they titled "A Synopsis of the Class Reptilia in Australia." On the masthead, Wells was listed as managing editor, Wellington as advertising sales manager. By publishing in a journal Wells controlled, the pair followed the precedent set by snake man Eric Worrell's publication of taxonomic descriptions in his own journal. As with Worrell's descriptions, the pair's decision to take this route to publication would look to many of their critics like self-dealing—in this case, criminally so.

In "Synopsis," Wells and Wellington named thirty-three new genera and fourteen new species, while either raising two hundred other species and eight other genera from the rank of subspecies or resurrecting them from synonymy—arguing, that is, that many of the names that Cogger and other taxonomists had deemed different names for the same species were in fact different names for different species. ("We here consider Macleay's *Diemenia angusticeps* as being a valid species of the *D. olivacea* complex," they wrote, "and take pleasure in formally resurrecting it.")

They seemed to delight in naming, calling creatures *Vaderscincus* ("named for Mr. Darth Vader"), *Eroticoscincus* ("erotic skink"), *Libernascincus* ("God of lustful enjoyment skink") *Licentia* ("freedom to do as one pleases"), *Phthanodon* ("to anticipate or do first"), *Cotundo* ("to crush"), *Anepischetos* ("unrestrained"), *Libertadictus* ("devoted to freedom"), *Solvonemesis* ("to set free the goddess of retributive justice").

Despite the rebelliousness these names hinted at, Wells and Wellington insisted that they acted only in the interests of conservation. "It is to be hoped that taxonomists will interpret our actions, not as anarchistic taxonomic vandalism, but as a decisive step intended to stir others into action," they wrote. "To us, the greater concern is that many species have their true identity masked by conservative taxonomic treatment, and are experiencing extensive loss of range under the misguided assumption that they are 'widespread and abundant' species."

As far as I can tell, this was the first use of that now-common phrase, "taxonomic vandalism."

~

In August 1984, some 180 herpetologists from Australia, New Zealand, and the Pacific islands gathered in Sydney for the second Australasian Herpetological Conference. They presented nearly a hundred papers and attended symposia including "Phylogeny of Elapid Snakes," "Rare and Endangered Snakes," and "Husbandry and Captive Breeding." In the *Herpetological Review*, conference organizer Rick Shine reported that the keynote address, "In Praise of Diversity," included "erotic films of tuataras coupling."

After the conference, a smaller group of herpetologists broke away for a second, smaller gathering. This group was known as the Australian Society of Herpetologists. Not to be confused with the similarly named but largely amateur Australian Herpetological Society, this group was composed mostly of professional scientists.

Among them the mood was cloudier. For months, word of Wells and Wellington's "Synopsis" had swirled through the Australian herpetological community. Details were scant—copies of the journal were hard to come by. "Rumor mills go crazy," says Shine, who was president of the professional society at that time. "It was clear there was massive angst. There were accusations of a lack of peer review. There were accusations of bad behavior, skullduggery. People were jumping up and down and getting terribly excited." Shine recalls taking in the scene with baffled detachment. "I'm an ecologist," he says. "I'd never had much reason to give too much thought to the names people apply to things. I'd been sort of aware that there was this rather religious reverence for names but I didn't really understand it."

Then an Australian Museum herpetologist who maintained ties with Wells showed up, bearing xeroxed copies of "Synopsis." As conference attendees began to fully absorb its contents, Shine says, many of them went "incandescent with rage."

In editorials, letters to the editor, and commentaries published in the following months, critics laid out their complaints. Wells and Wellington had offered only cursory descriptions of the new life-forms they described and often failed to give any justification for the ones they resurrected from synonymy. They appeared to have lifted type-specimen locations and numbers directly from Cogger's *Catalogue,* using that work's bibliography as the basis for much of their "Synopsis." They had made changes concerning life-forms that other scientists were already working on, perhaps in violation of the International Code of Zoological Nomenclature's nonbinding code of ethics and certainly in violation of good sportsmanship. They had made numerous grammatical and punctuation errors and had attached unbecoming and downright silly names to creatures, demonstrating an overall lack of seriousness. Everything they had done, they had done in unbelievable quantities. Their critics wondered why they had done it. What compelled them? Was it a sudden case of the mihi-itch? Or was their "Synopsis" exactly what they had denied it was, revenge, or an attempt to sow confusion? Was it, as Wells and Wellington themselves put it, an act of taxonomic vandalism?

Whatever the pair's motives, it was also apparent that they had satisfied all the important rules of nomenclature. "The names proposed," one herpetologist wrote, "clearly have legal status." Despite their low opinion of Wells and Wellington's work, other taxonomists would have to reckon with it—a task they estimated could take decades.

At the herpetological conference, attendees debated what to do. "I tried to defuse it as much as possible," Shine says. "It was completely futile." The crowd wanted blood. In his report on the conference, Shine wrote that officers of the Australian Society of Herpetologists had passed a series of motions. First, they noted that, despite a certain nomenclatural similarity, their organization had no affiliation with Wells's *Australian Journal of Herpetology*. Second, they affirmed their commitment to the rules and recommendations outlined in the International Code of Zoological Nomenclature, including its nonbinding code of ethics. Third, in a rare and dramatic step, they announced that they would soon send a petition to the International Commission on Zoological Nomenclature, the adjudicating body that rules on disputes concerning the Code. They planned to ask the commission to suppress Wells and Wellington's "Synopsis." Suppression was *damnatio memoriae*, erasure from the public memory—it would mean that taxonomists could forever ignore the "Synopsis" and all of its contents.

Undaunted, Wells and Wellington published a new edition of the *Australian Journal of Herpetology* in March 1985, a little more than a year after their first paper. It contained two articles, titled "A Classification of the Amphibia and Reptilia of Australia" and "A Synopsis of the Amphibia and Reptilia of New Zealand." In the two papers, they described 57 new genera and 146 new species and raised 9 genera and 110 species from synonymy. One commenter noted "a curious feature" of this second set of publications, which was that they contained citations for some five hundred papers written by Wells and

Wellington, published in *Australian Herpetologist*. "The bizarre aspect of these listed publications is that they do not exist," the commenter wrote. "*Australian Herpetologist* is a 'phantom periodical,' unknown to libraries." (Wells and Wellington insist that *Australian Herpetologist* was real, although they acknowledge that most issues consisted of little more than a typed-out version of a journal entry from one or the other of Wells's innumerable field trips; they acknowledge also that the periodical never achieved wide circulation.)

In the new taxonomic works, the pair only obliquely addressed the controversy their first paper had caused, writing that the "Synopsis" had "generated much comment from the herpetological community, many researchers agreeing with our fundamental thesis that the Australian reptile fauna was in need of taxonomic reappraisal." In the acknowledgments they thanked several particularly vocal critics for providing "the impetus to continue our interests in herpetology," and concluded with a series of epigrams, including a quote by Theodore Roosevelt: "It is not the critic who counts, nor the man who points out how the strong man stumbled, or where the doer of deeds could have done better. The credit belongs to the man who is actually in the arena; whose face is marred by dust and sweat and blood."

At this point, word of the shocking events in Australian herpetological taxonomy began to spread. Geoff Monteith, an entomologist and the curator of the Queensland Museum, learned of Wells and Wellington through a colleague, and acquired copies of their works. "The quality of W&W's taxonomy is unforgivably poor," he wrote in the August 1985 issue of the Australian Entomological Society's *Bulletin*, in which he recounted the herpetological saga. "The burden on conventional taxonomists to rectify the situation will be quite intolerable."

Monteith's article, titled "Terrorist Tactics in Taxonomy," caught the attention of scientific organizations around the world. "This may all seem to be very funny, and it's happening to zoologists, anyway, but there is no reason to be so smug about this," wrote a Dutch botanist, warning that a wayward plant enthusiast could make similar mischief. In the scientific journal *Nature*, a paleontologist predicted that "in the next few years, many of the world's biologists may find themselves beset by taxonomic problems originating from Australia." In his article, Monteith noted that at the end of Wells and Wellington's second treatise, the pair had mentioned a similar in-progress work devoted to New Guinea's reptiles, as well as one covering the reptiles and amphibians of the world. They intended to move from there into the Australian fishes and, after that, insects. "With their obvious determination," he wrote, "these aims are probably achievable."

In 1987, the *Bulletin of Zoological Nomenclature* published the Australian Society of Herpetologists' request that the International Commission on Zoological Nomenclature cast Wells and Wellington's publications into the "Official Index of Rejected and Invalid Works in Zoology"—a kind of black hole for names. The taxonomic and nomenclatural stability of Australian reptiles and amphibians "has been shattered," the society wrote. "If their three papers . . . are not suppressed, the effects on Australian herpetology will be devastating."

The request, designated ICZN Case 2531, drew more than eighty letters of support. But it also attracted criticism. The problem, as a group of French scientists pointed out, was that herpetologists' main complaint was about the quality of Wells and Wellington's taxonomic descriptions, not about the names they had used—it concerned the science of taxonomy, that is, not the nomenclatural filing system. For the commission to rule on a taxonomic question would set a frightful precedent, they wrote, impinging forever on the freedom of scientific thought. In a second letter, written in response to a herpetologist who had ignored Wells and Wellington's names and descriptions in a recent work, the French scientists wrote, "We are outraged by this attitude, which is best compared with the Stalinist falsification of history. . . . Will we be told that Wells and Wellington have simply never existed? Or perhaps they should be physically eliminated using an ice-pick?"

In December 1991, the commission delivered its decision. It criticized Wells and Wellington's conduct, writing that the pair had violated "virtually every tenet of the voluntary Code of Ethics." But it would do no more to censure them. Echoing the French scientists, it wrote that any problems with Wells and Wellington's papers were primarily a matter of taxonomy, not nomenclature. "The Commission has therefore decided it will not vote on this application," it concluded. "The case is therefore now closed."

Wells and Wellington could not be ignored, only reckoned with, and in the usual way: one species description at a time. To Australian herpetologists, it was a bitter disappointment. But their worst fears were never realized. Despite their stated plans, Wells and Wellington published no more treatises.

Wells, by most accounts, disappeared into a kind of hermitage in the mountains on New South Wales's North Coast. Wellington also remained on the North Coast. I visited his property east of Grafton and found him near the gate, looking at a death adder with his neighbor. I followed him up the long dirt switchback to his house. He wore fishscale-pattern swim trunks, flip-flops, and an off-white button-up shirt, partly undone. He was clean shaven with long, brown-gray hair in an unkempt ponytail, and seemed thin but wasn't quite. During one of our conversations, Wells had said his first impression of Wellington was of his "dashing blue eyes, and he'd stare at you, and you could tell by looking at his eyes, you know, he was really bright, very, very bright." And it was true. We sat under a gazebo, deep in the greenwood, and the sound of cicadas in the woods around us was loud as a lawnmower, and then later there was the sound of Wellington's partner, a Thai woman, up in the house pounding herbs with a pestle, and the air was thick and hot. The whole scene was strangely languorous, except Wellington's eyes. He stared at me intensely as he spoke. He was still angry about what happened.

He'd been surprised by the reaction to his and Wells's papers, he said. People called them terrorists and the AIDS of taxonomy. People called them thieves and idiots. "Which one was it?" he said.

"Was it crap and we were idiots, or were we correct and we'd stolen it?" Critics portrayed Wells as "this tin-pot undergraduate from the University of New England, this dumb schmuck undergrad who ridiculously started a journal and was out of his depth and then ran off and published his stuff," Wellington said. "He knows more than most professors." The critics were even more dismissive of Wellington—he was the advertising sales manager and high school teacher, little more than a typist playing lackey to Wells. The attacks were ad hominem, he said, focused on their credentials and methods to the exclusion of their work.

Finally he had enough. "I had all these headlines and media saying I'm a radical arsehole," he said. "It just became too stressful." He told Wells he didn't want to take part in any more taxonomic works. He taught high school for the rest of the decade, and then got a job in environmental education, taking children on trips into the bush, teaching them cultural heritage and edible plants. In the 1990s, he started working for the New South Wales National Parks and Wildlife Service, developing recovery plans for endangered species. The first plan he worked on was for the green and golden bell frog, the same frog he'd caught from a cistern when he was four years old. The once-common frog had become rare. It is now listed as endangered under state law.

After we sat for a few hours in the gazebo, Wellington took me on a tour of his property. Behind his house, we encountered a female wallaby that he said often hangs around. A joey was hidden in her pouch. She would not come closer to us, despite Wellington's attempts to entice her by tossing slices of bread. Against the wall there was a glass jar containing a dead snake floating in yellowish liquid. We drove in his red station wagon down to a clearing in a floodplain, then walked through tall grass to an overlook of the river that borders his property. He knew the trees and plants when I asked, both their common and scientific names, and said he'd inventoried 130 birds, 40 reptiles, 23 amphibians, and 28 mammals on his property.

The controversy over Wells and Wellington's papers irrevocably tied the men to each other. They were friends through birth and death and divorce. When Wells got married, Wellington was the best man. Each is wholehearted in his praise for the other. Wellington told me admiringly of Wells's vast property a few hours to the north, which he gained through a winning combination of strategy and chutzpah, and about the five-acre plots he rents to like-minded individuals, and about Wells's library, which has swelled to an Alexandrian scale. Wellington maintains that everything he and Wells did, they did in the interest of conservation. They recognized species that wouldn't otherwise have been named for years, for decades, perhaps ever, enabling people to talk about them, study them, conserve them. Almost certainly, they had saved creatures from extinction. The backlash, he concluded, was "sour grapes." As we stood overlooking a bend on the river, he told me there was a turtle in there nobody else knew about, still undescribed.

I had hoped to visit Wells at his sprawling compound and walk through his enormous library. Unfortunately, the timing didn't work due to his wife's familial commitments and other logistical complications, and so we arranged to meet instead at a café in Byron Bay, a surfing town on the North Coast. I arrived first and ordered a coffee. I sat at a table outside, hunched over and sweating, and watched the perambulations of beautiful and well-exercised people in mid-thigh swim trunks and bikinis who possessed thick and lustrous hair, who wore either flip-flops or no shoes at all to mediate between their soles and the blessed Earth. I could tell from miles off that their teeth were perfect, these people, and they were good in bed, and they were maybe into crystals but not to an inappropriate degree.

Scattered among these tanned Apollos and Aphrodites were older people, weathered but still taut and golden, the kinds of elders who appear in commercials. I studied them closely, wondering if maybe one would turn out to be Wells, but then I saw him and he was unmistakable. He was so unmistakable that I wondered if anybody else could see him. He was dressed head to toe in battle-stained khaki. His boots were in a state of near-total dissolution. His head was orbited by an enormous corona of translucent yellow-white hair. More yellow-white hair erupted from the schism in his shirt and covered his densely freckled arms, hands, and fingers. He had a mustache, hazel eyes that twinkled, actually twinkled, and a sunburned forehead that was starting to peel. I could smell him before he reached the table.

Almost immediately after sitting down, he informed me that, according to both astrophysics and his own premonitory powers, and due partly to the cycle of solar flares and partly to Earth's unfortunate proximity to the Sagittarian galactic core, the planet's magnetic poles are in the process of reversing. This is something that happens with some regularity, geologically speaking, and it was already wreaking havoc with the planet's living inhabitants, causing them to behave in all kinds of unusual and pestiferous ways. When this reversal really gets going, though, he said, it will cause Earth's ballast to shift in the hold, tipping the entire planet on its keel, sending the oceans sloshing over the decks, and inflicting unspeakable damage. Whatever remnants of humanity survive this ordeal will be plunged immediately back into the Iron Age, he told me, from which point they will follow a slow and inexorable decay back to the Stone Age.

He could see all of this, he said. "I see it clear as day." He and his tenants were busy building bunkers. The community they've formed is fully self-sufficient and located in what he estimated to be "probably the best place on the planet to withstand what's coming." It is in this postdiluvian future that his library will be fully realized. The internet, he said, "is going to be like a dim, distant glitch." Whatever knowledge survives will be what is written down. He was coordinating with nine other like-minded people scattered around the world, who together had gathered nearly a hundred million books. Wells estimated his share to be roughly seventeen million books.

Considering our library-visiting Cro-Magnon heirs, I thought of H. G. Wells's *The Time Machine*, in which, far in the future, the time traveler finds a library in the Palace of Green Porcelain. The library is moldering and also infested with morlocks. Coincidentally, Richard

Wells told me that H. G. Wells was a distant relation, and he recalled that when he was a young man he encountered a picture of a young H. G. standing beside the skeleton of a gorilla and was shocked to find that the young H. G. and the young Richard were indistinguishable. It wasn't merely a resemblance—it was a picture of himself.

This, admittedly, was not the conversation I had expected, but I found that despite it not conforming to my ordinarily literal understanding of reality, everything Wells was telling me made perfect sense. I sucked my coffee and scribbled notes and thought that Wells was perhaps the most intelligent and charismatic person I'd ever met. As he told me about a secret and benevolent council of extrawise people who help steer the course of civilization, I considered whether I ought to abandon everything I held dear and start paying rent on one of his five-acre plots.

Still, when he paused momentarily to take in air, I forced myself to ask him about the Wells and Wellington Affair. This was almost an afterthought, given his more recent activities, but he indulged me. When the papers were published, he was surprised by the backlash. The former professors who had advised him on the *Australian Journal of Herpetology* disavowed him, he was voted out of his post as president of the Australian Herpetological Society, and then soon after his father died. "My dad was dead, and I was still looking after my mum, and I was in no mood to argue with academics," he said.

In the decades that followed, Wells worked a hundred different jobs and visited every place named and unnamed, and mostly he disappeared from public view, occasionally surfacing at reptile conferences. Wellington sent me a picture taken in 2019 of him and Wells at the Australian Herpetological Society's seventieth-anniversary dinner. They were sitting at a table with Hal Cogger, the former Australian Museum curator, whose *Catalogue* had motivated and allegedly aided Wells and Wellington in their own taxonomic efforts. In the years since the Wells and Wellington Affair, Cogger's *Reptiles and Amphibians of Australia* has run to seven editions. Each edition has contained more of the creatures Wells and Wellington named.

"Look, I need to make one point quite clear," Cogger said when I asked about this. "And that is, Richard and Ross are extremely knowledgeable herpetologists. They know the fauna well, they've done lots of field work, they've read a lot, so I would never question their knowledge of herpetology. Some of their actions in relation to the *Catalogue* and other actions they've taken, I find . . . not really very nice sort of stuff. But there's no issue as far as I'm concerned about their knowledge of fauna." Over the years, as taxonomists have picked through Wells and Wellington's work, animal by animal, description by description, they've rejected some but accepted others. "Many of the changes proposed by Wells and Wellington have been substantiated by later, evidence-based studies," a group of researchers wrote in 2006. "Both are highly skilled herpetologists, particularly with snakes and lizards," wrote another scientist. As another researcher put it, "History has to some extent vindicated Wells and Wellington."

The more forgiving view of Wells and Wellington may be due in part to recent events that have given the scope of their activities a new context. Wells brought it up, unprompted. "Ray Hoser saw what Ross and I had done and he tried to emulate us," he told me. "Hoser is the devil's spawn. He saw what we did and he thought that he could do it so easily."

CHAPTER 7

WHISTLEBLOWER

My own responsibility, however, as it has been throughout my writing of this entire narrative, is simply to record whatever I may be told by my sources. . . . I know their names but do not record them.

—HERODOTUS, *THE HISTORIES*

On the morning of what would turn out to be a very busy day, Raymond Hoser picked me up from a café in the northeastern suburbs of Melbourne. He was fifteen minutes late, moving fast, but he slowed his black Nissan Micra enough that I was able to get nearly seated before he rocketed back up the street.

He was wearing shorts, slip-on black leather boots, and no shirt. He had an earpiece in his right ear and was talking with a potential client about what was possibly a snake in their house. Without pausing the conversation, he reached over and shook my hand with a constricting grip, locked his pale green eyes on mine, and flashed a wide smile that didn't show his upper teeth. He was clean shaven, mostly bald, powerfully built, with meaty hands and a shallow philtrum and a nose that ended in a hemispheric bulb. Even in the relative stillness of the car, he exuded a caged, teetering energy.

On the dashboard was a faded sticker from an elementary school that read: "Contractor: Raymond Hoser, Snake Man. Reason for Visit: Snake." Below the sticker was a stack of four cellphones. The car smelled strongly of sweat and an ammonia tinge that I would soon come to recognize as the scent of reptile feces. The rear two-thirds of the car was filled with clear plastic tubs that contained, Hoser told me later, twelve pythons, eleven lizards, six turtles, six frogs, four venomous snakes, and a crocodile named Joshua, who was in the tub near my elbow. Her mouth was ajar and she was looking right at me.

When Hoser hung up the phone, he told me he was running late because of matters related to an ongoing legal case involving several Australian and foreign turtle taxonomists, who he said were members of the Wüster Gang and who refused to use his turtle names. To understand it all, you'd have to go back many years, even centuries. "Everything I tell you is documented in the various papers I've written," he said. "Have you seen these papers? They have everything in there. Got everything in black and white, got the accurate story. Then you go to 2000, and . . . Fuck!"

He realized we were going the wrong way. "Eh, I'm going to do a U-turn," he said. "Fuck. We're going to be late now. What time is it? Fuck, I've got that one wrong. That's all right, we'll survive. I'll just ring her and tell her we'll be fifteen minutes late." He pulled out a small notebook and began flipping through it, still weaving through traffic. "I'm not God," he said to himself. "I'm not a miracle worker. Oh shit, that was bad. I thought I was going to fucking Scoresby. We'll survive. It's only one party. Bear with me." Glancing back, he whipped the Nissan Micra across several lanes of traffic, careened up and over a curb, then accelerated back in the other direction. "Ah, okay, we're off and running," he said. "That was bad planning. Now we're fifteen minutes late." He paused and considered. "It's a kids' party. Where the fuck are they gonna go?"

As we merged onto a westbound highway, Hoser noted an empty five-gallon bucket beside the road. He made a phone call to the woman whose child's birthday we were headed to, telling her we were forty minutes away, then entered the address in Google Maps, which showed we were fifty-five minutes away. He explained the nature of his navigational mistake, which involved a canceled party in Scoresby, the location of which he still had in mind, and he said he was nearly never late to birthday parties. The good news, he said, was that there were no more parties today. When it was over, we could go back to his house and clean reptile cages, which ordinarily he would have cleaned before, but hadn't due to his legal commitments and because he had more parties the next day. Then he paused, seeming to perceive me fully for the first time. "Do you have much experience with reptiles?" he asked.

I told him I'd had a garter snake when I was a boy.

"That's it?" he said.

~

No snake man in Australia and perhaps even in the world—not Charles Underwood or Joseph Shires, not Eric Worrell, not Wells or Wellington—has attracted more attention from reporters and chroniclers than Raymond Hoser. The first of his innumerable appearances in the press came in 1973, in a *Sydney Morning Herald* article headlined "Hard Boiled Gamblers' Sweet Win." The story was about a turtle track that Raymond's father, Len Hoser, had built in the backyard of their home in Greenwich, a Sydney suburb. Len told the reporter that Raymond, age ten, and his brother Phillip, thirteen, had collected the turtles from a creek near their house and reared them on mincemeat. Len had imposed rules to promote sportsmanship and the turtles' wellness: no tapping of the turtles; no redirecting turtles that turned around mid-race. "And there's no doping of course," he said. An accompanying photograph showed the Hoser boys leaning over a six-lane track, which the article reported had individual stables, starting gates, and a betting table. Phillip, in clear violation of track rules, was reaching over to assist a laggard in lane 1. Raymond's gaze, meanwhile, was fixed on a pair of turtles in lanes 3 and 4, an inch off the finish line. He was grinning in sheer delight.

The family had moved from England five years earlier. Hoser told me that he began catching reptiles almost immediately. "I was catching snakes straightaway. I was catching adders when I was four years old," he said. "All the girls thought I was batshit." Like Worrell, like Wells and Wellington, Hoser had a dream for as long as he can remember. "I had my whole life trajectory worked out," he said. "Go to Sydney University, become a professor, make lots of world-shattering discoveries."

But just a year after the *Sydney Morning Herald* turtle-racing article, New South Wales passed its National Parks and Wildlife Act, the law that forced Richard Wells to turn over his reptile specimens to the Australian Museum and helped drive Eric Worrell from his reptile park. For Hoser, the law was perhaps even more incisive. The New South Wales Parks and Wildlife Service started "grabbing people off the street, locking people up for having blue-tongues," he said. "Schoolkids were going to jail for keeping a lizard. You'd go to jail for catching a tadpole in your backyard." Being precocious and clearly gifted in his herpetological pursuits, Hoser soon caught the attention of the wildlife service. "I was a standout, even at that young age, getting raided and tied to chairs at fourteen years old," he said. These and other details were difficult to corroborate independently, due to the nature of the Australian legal system, which prioritizes individual privacy over public access to records, and due to the length of time that has elapsed since purported events, and also due to the general reluctance of Australians I spoke with to commit their memories of Hoser to record.

Richard Wells was one exception. He met Hoser at a meeting of the Australian Herpetological Society during his time as society president. Hoser was in his early teens. He invited Wells to see his reptiles, so Wells went to Hoser's house and met his parents and saw his collection of reptiles, which had grown beyond just turtles. Later, Wells watched Hoser handling venomous snakes in a parking lot. "He was a bit scared of them, the snakes, at the time," Wells recalls. But Wells was nevertheless impressed. "He still really showed me that he had great knowledge." Wells noticed another of Hoser's qualities, which was that Hoser did not seem to appreciate the 1974 law's proscriptions, or really any other proscriptions, for that matter. Hoser, it was already clear, would sooner be an outlaw than bow to laws he thought unjust.

Hoser graduated from high school and, as he had dreamed, entered Sydney University. The following year, in a letter to the editor published in *The Herptile*, an amateur herpetological journal based in England, the teenage Hoser displayed early signs of his future career as a whistleblower. Many of his Australian amateur herpetologist peers, he wrote, kept reptiles in "substandard conditions, knowing that when their reptiles die or escape, they can be replaced without too much difficulty." He also hinted at his own scientific activities, writing that he was involved in "detailed research on all aspects of the biology of Death adders," politely concluding, "I hope that I haven't been long-winded, and trust that this letter has enlightened *Herptile* readers."

Over the next two years, *The Herptile* published a number of his articles, on Australian pythons, on the frequency of skin-shedding in captive Australian snakes, on the role of pelvic spurs, and on other topics. But the Wildlife Service worked to stifle his budding scientific career, Hoser says. In 1982, *The Sydney Morning Herald* published a short article about Hoser, then twenty years old. He had been charged with keeping five death adders and three pythons without a license. His attorney told the judge that the police sometimes called Hoser to remove snakes found on private property, and that he wanted to "continue collecting, breeding, tabulating and doing scientific work with reptiles, and he hoped to make a career of it." The judge dismissed the charge, ordering Hoser to pay sixty-three dollars and turn over his snakes to the Taronga Zoo.

Just over a year later, the paper published another article about him. Residents of Hoser's Redfern neighborhood called the police to kill a python that had appeared on the streets. When the police discovered three more free-roaming snakes, all apparently escaped pets, his neighbors told police they should talk to Hoser, "the proud owner of 52 snakes," including 33 death adders. Hoser told police he'd been on holiday when the snakes turned up and they weren't his. Someone must have planted them, he told the reporter. "I found a python in my bed when I came back," he said, "but it wasn't one of mine."

A month later, the National Parks and Wildlife Service raided his home, seized his snakes, and charged him with five counts of breaching the National Parks and Wildlife Act. A year after that, in 1984, the *Herald* published yet another article about Hoser, whom it referred to as "the Snakeman of Redfern." Police had raided his home again, this time discovering more than fifty live snakes, plus hundreds of dead snakes and other reptiles frozen or preserved in pickle jars. "As a result of the raid on his home by NPWS officers," the newspaper reported, "the Sydney City Council became concerned with the man's infatuation with snakes."

~

The old newspaper articles were helpful, Hoser told me as we careened across Melbourne. They helped him establish that for decades he had been known as not only a snake man but *the* Snakeman, which helped him persuade the trademark authorities that he had the exclusive right to "snakeman," "snake busters," and other terms. He explained that, by providing him with grounds to sue, his twenty-odd trademarks were what allowed him to overcome his enemies' attempts to ruin his results on Google and other internet search engines through a practice known as "negative search-engine optimization." These enemies, who he said deprived him of business and sullied his reputation by disappointing the public with their suboptimal snake shows, appeared to be ordinary snake catchers and snake performers, but were in fact working with the German snake-venom expert Wolfgang Wüster and the other taxonomists and academics who comprised the Wüster Gang. "Your enemy's enemy is your friend," he said. "When they decide to attack me, they see who else has got an ax to grind against me and they join forces."

But all of that—the trademarks and questions of search-engine optimization and the epic battle with the Wüster Gang—was still far in the future. Soon after the 1984 raid on his home, Hoser says, he received poor grades in his classes at Sydney University, undeservingly, he thought, so he went to see the vice chancellor. "He says, 'Raymond, you've been in the news about Parks and Wildlife,'" he recounts. "'They've come in here and said we've got to stop you getting a degree, we can't let you pass.'"

Hoser marks this as a low point. He had no degree, was in debt, and felt weary from the Wildlife Service's endless persecution. For the first time since he was a boy, he had no reptiles. "I gotta get out of Eden," he recalled thinking. In 1985, he moved to Melbourne. There, he began driving a taxi, he says, working seven days a week, sleeping in a shared house. In what little spare time he had, he worked on his first book. Published in 1989, it was titled *Australian Reptiles and Frogs*. The book included more than six hundred photos, which, according to the book's cover, was "the largest collection ever published," with "world first" images of interspecies snake copulation and snakes "feeding and copulating simultaneously," as well as pictures of mid-metamorphosis tadpoles, a death adder with an engorged hemipene, and a quadriptych arrangement of a tiger snake eating a rat. The book had sections on herpetological husbandry, breeding, acquisition, conservation, and photography, as well as a legal advisory: "In some states," he wrote, "otherwise law-abiding citizens prefer to hold reptiles and frogs without the knowledge of wildlife authorities. This is because of actions taken by the authorities against reptile license-holders, including break-ins and theft of specimens." The real bandits, in other words, were wearing badges.

Hoser repeated these allegations in his second book, *Endangered Animals of Australia*, published in 1991, when he was twenty-nine. The biographical sketch on the book's back cover described him as an "internationally known authority on Australian wildlife," an early proponent of whale conservation, and developer of "some of the most impressive reptile captive breeding facilities in this country." Nearly all smuggling of Australian wildlife, he wrote, "relies on the corruption of government and law enforcement officials," who often stole animals from registered reptile owners. The section on smuggling originally stretched to twenty thousand words, he says, but the publisher cut it. This material became the seed of his third book, *Smuggled*.

~

Wildlife smuggling, on the one hand, presents a series of uncomfortable and perhaps unanswerable questions about the human psyche, about our greed and covetousness and insatiable appetite not just for novelty but for rarity itself, all concerning animals that would almost certainly choose a life in the wild. On the other hand, it seems quite simple: When demand is high and supply low, prices will rise. So it went in Australia, as legislators across the country enacted laws to protect native wildlife, squeezing supply while doing little to dampen demand. Through the 1970s and '80s, newspapers reported hints of the growing trade. A man was arrested at the Perth airport with a suitcase containing 39 birds, half of which were dead. A truck seized at Derby contained 361 birds, many of them also dead. Two men were arrested in Bordertown, South Australia, for possession of twelve wombats. Two men were arrested while attempting to smuggle fourteen parrots to Europe; they were believed to be connected to a criminal mastermind known as "Mr. Bird-Brain." A Dutchman was intercepted at the Melbourne airport with four diamond pythons, four Children's pythons, one gold crown snake, four red-bellied black snakes, and five blue-tongue lizards. Reptiles were discovered concealed inside postal packages, Christmas ornaments, toys, books, socks, stockings, cigarette packets, and a specially built vest. A sulphur-crested cockatoo was reported to cost $1,500 on foreign black markets; a Major Mitchell's cockatoo, $3,000; gang-gang cockatoos, $20,000 per pair; a glossy black cockatoo, $50,000. A green tree python cost $1,500. A blue-tongue lizard cost $2,000. Carpet pythons were reportedly going for $20 a foot. In Coffs Harbour, thieves broke into a wildlife sanctuary and stole twenty-three reptiles, including death adders, tiger snakes, and a twelve-foot amethystine python. "This is obviously the work of a professional snake man," said Mrs. Mutton, wife of the sanctuary manager. "Money holds no object to snake men—they would sell their soul for a good reptile."

In a 1992 report, Boronia Halstead, research officer at the Australian Institute of Criminology, traced the path of wildlife smugglers from airstrips in northern Australia to Thailand, Indonesia, and Singapore, where animals were paired with forged documents before traveling on to buyers in the United States, Europe, and Japan. Halstead noted the rising alarm of national and international lawmakers, the adoption of international treaties meant to suppress the wildlife trade, and the increasing focus on wildlife smuggling by various branches of Australian law enforcement. Only in passing did she mention the possibility of corruption—that some of the people supposedly cracking down on smuggling were smugglers themselves.

This was a significant oversight, according to Hoser. Both the former director of Sydney's Taronga Zoo and its former head reptile keeper "were well-known wildlife traffickers," he announced on the opening page of *Smuggled*, published in 1993, just a year after Halstead's report. So were countless members of the wildlife services and customs. The American Mafia was also involved. While Halstead and other observers seemed to assume that most smuggled animals were collected from the wild, Hoser argued that the main source was, in fact, licensed private keepers like him, whose animals were confiscated by corrupt officials during raids and then resold. "The only people who know who holds what wildlife and where, are the licensing and law enforcement officials in these government departments," he wrote. "Around the world as statutory protection of wildlife has become more widespread, it is these officials who have become an important link in all major smuggling rackets." He had been investigating this pervasive official corruption since his youth, he continued, and had obtained much information from recordings "made by me or others, often when one or more parties to the conversation were unaware that they were being taped."

One of these conversations was with Steve Gordon, the New South Wales western region commissioner for the Aboriginal and

Torres Strait Islander Commission. Gordon alleged that New South Wales wildlife officials had taken bribes from kangaroo-meat processors to suppress their competition, Hoser wrote in *Smuggled*. Gordon also claimed (per Hoser) that he'd witnessed certain kangaroo-meat processors feed the remains of a missing kangaroo-meat processing boss, Andy Komarnicki, aka "the Kangaroo King," through a meat grinder. In a purported transcript of the conversation, Gordon allegedly told Hoser: "They were mixing it up with other meat, see."

Soon after *Smuggled* was published, the New South Wales National Parks and Wildlife Service sent a letter to the publisher, Apollo Books, writing that the book was "highly defamatory," and demanding that all copies be withdrawn from sale. In Australia, the opposition in Parliament called for a federal inquiry into Hoser's allegations of corruption among officials, which an opposition spokesperson called "very disturbing." In the United States, wildlife advocates issued similar calls. Soon after, the Australian minister for the environment asked the Independent Commission Against Corruption to investigate the alleged involvement of the National Parks and Wildlife Service in trafficking and asked the service to withdraw its threat of legal action against Apollo Books. Six months later, the service issued a statement reporting that the corruption commission had cleared it of wrongdoing. Hoser was undaunted. "The cases I have included in the book are the tip of the iceberg," he had told a reporter soon after *Smuggled* was published. "I could fill dozens of books with the information I have."

The following years passed in a blur of activity. In 1995, Hoser published *The Hoser Files: The Fight Against Entrenched Official Corruption*. In 1996, he published *Smuggled-2: Wildlife Trafficking, Crime and Corruption in Australia*. In 1999, he published *Victoria Police Corruption* and *Victoria Police Corruption-2*. The inside back cover of the latter book shows a full-page photograph of a dead man's face with his eye blown out. Hoser alleged that the man was murdered by Victoria police. The placement of the graphic image was meant to hook readers, he told me, and noted disapprovingly that my first book, which I'd mailed to him months earlier, included no pictures.

When reporters and reviewers notice Hoser, as they frequently do, but don't take time to absorb the breadth of his story, they often seem unsure what to make of him. A reviewer in the *Australian Rationalist* deemed Hoser a "tireless investigator and chronicler of official corruption" and applauded his efforts, but noted that his books were written "in a very haphazard style that makes them difficult to read. . . . The overall effect is an amazing mess in which you may sense that there is important information to be found, but cannot possibly glean it. The lack of references is also worrying."

Police and officials responded to Hoser's whistleblowing with repeated defamation suits and, he says, by harassing and arresting him on trumped-up charges. On his website, smuggled.com, Hoser lists dozens of key events, including:

19 July 1992: Hoser's car found stripped adjacent to Northcote Police Station. Police tell Hoser "It serves you right." . . .

29 November 1993: Hoser pulled up for going through malfunctioning lights at King Street, Melbourne. . . .

11 January 1994: Hoser beats another of many traffic charges in court. Police officers . . . taped committing perjury. In line with common practice magistrate . . . and Police refuse to charge the officers. . . .

24 January 1994: Hoser sends to Vicroads two identical letters inquiring re malfunctioning traffic lights. . . .

17 February 1994: Hoser appeal for traffic light matter in front of Susan Blashki. As usual Police side took steps to ensure proceedings not taped. Also as usual corrupt Judge . . . sided with Police and didn't want matter taped. As usual Hoser still had matter taped. . . .

18 February 1994: Hoser house raided by police. Carloads of material taken including tapes of Balmford hearing and all relevant documents. Thousands of other tapes, computer disks, documents, etc, taken. House trashed. . . .

3 October 1994: Policeman . . . attempts to extort money from Hoser. . . .

1 April 1995: Hoser arrested by Police in South Melbourne and falsely accused of stealing his own taxi. He is then brutally bashed and thrown out of Police station. Admitted to Royal Melbourne Hospital. . . .

3 October 1995: Hoser 'convicted' of perjury. Tape of Hoser's 28 minutes of evidence . . . was deliberately kept away from the jury by the prosecution side and judge.

4 October 1995: Hoser jailed for a minimum of four months as a result of the above conviction. . . .

26 March 1997: Talbot traffic charges against Hoser thrown out by Magistrate. . . . As usual for magistrates . . . was not interested in taking action against Police for proven perjury during the case.

23 April 1997: Appeal against . . . conviction dismissed. Hoser jailed for four months.

Throughout this period, Hoser continued working as a taxi driver, and often argued publicly for what he viewed as the industry's interests. Melbourne "is one of the few cities in the world stupid enough" to have trams "sitting in the streets instead of the museums where they belong," he wrote in one letter to the editor. "As a full-time professional taxi driver, I know of innumerable safe drivers who have been stripped of their licenses by cowboy police with radars and cameras after being booked for breaking inappropriate speed limits at no risk to anyone," he wrote in another. "Taxi drivers believe their industry should be run by someone with recent taxi driving experience and not a career bureaucrat," he wrote in a third, which he signed "Raymond Hoser, Taxi Drivers' Association of Victoria." Three days later, Walter Connolly, president of the Taxi Drivers' Association of Victoria, published his own letter to the editor. It read, in its entirety: "The Taxi Drivers' Association of Victoria has nothing to do with Raymond Hoser in any shape or form."

In 1999, Hoser ran for the Frankston East seat on the Victoria Legislative Assembly. During the campaign, he sat for an interview with morning-show journalist Steve Price. "And just finally, you've got a criminal record," Price said toward the end of the segment. "Do you think you should let the people of Frankston East know the details of that?"

"I have been charged with criminal offenses and they are public in my books, which have sold thirteen thousand copies," Hoser said. "There is no secret, the charges only were laid against me after I blew the lid on corruptions in government."

"You've been charged with assault[ing] police," Price said.

"That was overturned on appeal," Hoser said.

"And you've been charged with a dog attack," Price said.

"Sorry?" Hoser said.

"A dog attack on a person," Price said.

"No, that's not true," Hoser said. "You're wrong, I'm sorry about that."

"Prahran Magistrate's Court 1998," Price said. "Convicted and fined two hundred dollars."

"No, you're wrong on that, that simply hasn't occurred," Hoser said.

"Perjury," Price said.

"No, wrong again," Hoser said.

"Six months with a two months suspended sentence," Price said.

"No," Hoser said, "I would suggest that the actual criminal record . . . that you are misquoting, are the charges which I did in fact beat, are in fact public in one of my books, *Victoria Police Corruption-2*—"

"Assault by kicking," Price said.

"If you don't cut me off, please," Hoser said.

"Behav[ing] in an offensive manner, aggravated littering, another case of behaving in offensive manner, assault on police—it's a pretty long criminal record," Price said.

"I think you've got your facts wrong," Hoser said.

"Well, I have a copy of your police record," Price said; "I don't see that I can have it wrong."

"Well, the record you have is incorrect," Hoser said.

Hoser subsequently wrote an article about the interview that was published by *Crikey*, an Australian online magazine. Price deemed the contents of this article defamatory, and sued *Crikey*'s publisher, Stephen Mayne. Mayne lost and was forced to sell his house to settle the damages. Hoser placed last in the election, winning eleven votes out of more than twenty-five thousand cast.

~

During some of my time in Melbourne, Hoser was too busy with his legal efforts to have me around, so I spent several days lurking at the State Library Victoria. When I got tired of poring over important books and documents, every twenty minutes or so, I would drift downstairs to the South Rotunda, where Ned Kelly's armor was on display.

The son of an Irish convict, Kelly was in trouble with the law from a young age, eventually leading a band of outlaws on an eighteen-month spree of robbing banks, destroying telegraph poles, and taking hostages, culminating in a shoot-out with the police that began in the early morning hours of June 28, 1880, near the town of Glenrowan, Victoria. During the showdown, Kelly and three other members of the gang wore suits of armor made from plows. Now mounted in a glass case, Kelly's armor is like a child's drawing of a medieval knight, oddly angled, asymmetrical, a hash of rivets, bolts, straps, and wires, a floating torso of black oxidized steel topped by a tin-can helmet with a narrow eye slit. It is pocked with dents where bullets bounced. By the time he was captured, Kelly had been hit eighteen times in the armor and sustained dozens of wounds on the unprotected parts of his body.

Throughout his time on the lam, during which he killed four men, including three police officers, Kelly cast himself as part of a grand struggle: between Australia's landed aristocracy, or "squatters," and itinerant farmers, or "selectors"; between colonized Irish and the oppressing "Saxon" Crown; between corrupt police and honest outlaws. "He saw himself as a Robin Hood leading poor, suffering rural workers against a vicious ruling class," wrote Keith Dunstan in his book *Saint Ned*. He quoted Clive Turnbull, who wrote that the public "saw in Kelly those qualities which are deemed the most desirable in the Australian conception of manhood—courage, resolution, independence, loyalty, chivalry, sympathy with the poor and ill-used." After the court sentenced Kelly to death on November 3, crowds gathered in protest. "There were about 4,000 inside and a large concourse outside the building," one reporter wrote of a protest in Melbourne organized by Kelly's lawyer. "A resolution was unanimously carried in favor of a petition to the Government for a reprieve, on the ground that Kelly was not guilty of murder, and that the police came out to shoot him and his brother, and that he did not intend shooting the police but only to make them prisoners and take their firearms away."

Kelly's purposeful presentation of himself in this manner and the public's ready acceptance of it point to how deeply the bandit-hero is embedded in the collective consciousness. Kelly was playing a character any audience would recognize. As Paul Angiolillo writes in *A Criminal as Hero*, there have been Robin Hoods in every culture, in every era. "The forest may be replaced by the city slum or the western plains," he writes, "the bow and arrow or musket, by the machine gun or six-shooter; but essentially the myth remains the same: the fearless, independent outlaw, dedicated more or less to the defense of the helpless, the rightings of wrongs, the humbling of the rich and powerful, and the dauntless display of extraordinary courage, deemed to be beyond the ken of the common run of men."

In a world of compromise and nuance, the bandit-hero offers simplicity. "He has no practical scheme for improving the human condition," writes historian J. C. Holt in *Robin Hood*. "There is no sense at

any stage of the legend's history that man's lot might be improved by
the sweat of his brow. Robin simply intervenes. . . . Robin belongs to
that world of heroes and villains in which the heroes are so supreme,
so magical in their mastery that they will always win."

But referencing this character and living it are two different things. Kelly's deeds and words stand in contrast to his heroic
pretensions. In the midst of his final spree, he dictated a letter to a
member of his gang. The lengthy document is not quite an autobi-
ography, not quite a political screed, but something in between, a
Joycean stream-of-consciousness account of his various run-ins with
the police and his fellow citizenry, ending with a public ultimatum.
It read, in part:

> She was a chestnut mare, white face, docked tail, very remarkable.
> Branded as plain as the hands on a town clock. The property of
> a Telegraph master in Mansfield, he lost her on the 6th, gazetted
> her on the 12th of March and I was a prisoner in Beechworth
> Gaol until the 29th March. Therefore I could not have stole the
> mare. . . .

> I was blamed for stealing this bull from James Whitty Boggy
> Creek. . . . Not long afterwards I heard again I was blamed for
> stealing a mob of calves from Whitty and Farrell which I knew
> nothing about. . . .

> . . . It seems that the jury were well chosen by the police as there
> was a discharged sergeant amongst them which is contrary to
> law. They thought it impossible for a policeman to swear a lie
> but I can assure them it is by that means and hiring cads they get
> promotion. . . .

> . . . But they knew well I was not there or I would have scattered
> their blood and brains like rain. I would manure the Eleven Mile
> with their bloated carcasses and yet remember there is not one
> drop of murderous blood in my veins. . . .

. . . But as for handcuffing Kennedy to a tree or cutting his ear off or brutally treating any of them is a falsehood. If Kennedy's ear was cut off, it was not done by me. . . .

. . . And any person aiding or harbouring or assisting the police in any way whatever or employing any person whom they know to be a detective, or cad or those who would be so depraved as to take blood money, will be outlawed and declared unfit to be allowed human burial. Their property either consumed or confiscated and them and theirs and all belonging to them exterminated off the face of the earth. . . .

. . . And do not attempt to reside in Victoria but as short a time as possible after reading this notice, neglect this and abide by the consequence which shall be worse than rust in wheat in Victoria or the drought of a dry season to the grasshoppers in New South Wales. I do not wish to give the order full force without giving timely warning but I am a Widow's Son, outlawed and my orders must be obeyed.

They are the words of a "violent and vindictive man who demonstrated prominent psychopathic features including pathological lying, callous lack of empathy for others and a parasitic lifestyle," as psychologists Russ Scott and Ian MacFarlane put it in a 2014 analysis. "Kelly's own florid diatribes and correspondence further illuminate his grandiose sense of self-worth and his inability to accept responsibility for his criminality." So too with Robin Hood. Far from the merry, mischievous, Roger Miller–whistling prankster of Howard Pyle and Disney, early poems portray him as delighting in violence, killing "bloodily and with zest," Holt writes in *Robin Hood*. "He not only shoots the sheriff but beheads him."

Such is the cultural power of the bandit-hero that it can cloud in subjectivity and nuance deeds that would otherwise read as plainly bad. This type of moral obfuscation has been successfully employed by such figures as the Unabomber, Osama bin Laden, and the wealthy young man who recently murdered a health-care executive in Manhattan, describing his act as a blow against the iniquities of an entire industry. The bandit-hero, in other words, is a costume, universally understood, one that anyone feeling cast out or mistreated or ignored might reach for.

~

Even as Hoser drove a taxi and became a prolific whistleblowing author, he continued his scientific studies. Throughout the late 1980s and '90s, he contributed steadily to the herpetological literature. There were articles on the feeding behavior of common scaly-foots, on the immunity of snakes to their own venom, on snakes swallowing their own teeth, on hybridization among three species of Australian pythons, and on sperm storage in the blue-tongue skink. In these articles, he took a markedly different tone than in his whistleblowing works. As a whistleblower, he was aggrieved, rebellious, the righteous outcast decrying a corrupt system. The writing often bordered on incomprehensible. As a scientist, by contrast, he wrote soberly, clearly, with an obvious grasp of the relevant concepts and jargon. He was still trying to fit in.

Also apparent during this period was Hoser's continued affection for death adders. He wrote articles about the frequency of death adder shedding, the Mendelian genetics of death adders from the West Head area of Sydney, and the mating behavior of death adders, including notes on the sexual attractiveness of death adders, and about a death adder he observed eating an inappropriately large lizard. In a 1995 article, he wrote that one of his death adders had envenomated him while he was measuring it and "stuck my finger down his throat by accident." As a "precautionary measure," he checked into a hospital. His symptoms peaked about twenty hours after the bite. "By that stage I was barely conscious or able to speak or do much of anything else," he wrote. His speech was an unintelligible slur, he could barely open his eyes, and his pulse slowed. Monitors found that airflow through his esophagus fell to just 10 percent of normal. Through it all, he wrote, his "mind was still 100 percent."

In the same article, Hoser offered a lengthy history of death adder taxonomy. In 1863, Günther argued that certain death adders of New Guinea and Indonesia were their own species, which he called *Acanthophis ceramensis*. In 1877, Ramsay named the death adders of northern Australia *A. praelongus*. Macleay described *A. laevis* in 1878, Boulenger described *A. pyrrhus* in 1898, and Loveridge described *A. rugosus* in 1948. In the 1980s, in their two papers on Australian herpetology, Wells and Wellington named three additional species, *A. armstrongi*, *A. hawkei*, and *A. lancasteri*. In his article, Hoser hewed to the herpetological mainstream, following Hal Cogger and other contemporary authorities in recognizing just three valid death adder species: *A. antarcticus*, *A. praelongus*, and *A. pyrrhus*.

Hoser noted, though, that there was likely another valid species of death adder, native to the Pilbara Desert of Western Australia. In 1992, a photo in the *Australasian Post* had shown Western Australian Museum curator Ken Aplin holding a death adder close to his face. The article called it a "red death adder," and said that the "brilliantly-colored snake—patterned with bold black stripes and jet black head," had been discovered that year. "It's rare to find a large, brightly colored snake like this that hasn't yet been collected and described," Aplin told the magazine. The next year, *Monitor*, the journal of the Victorian Herpetological Society, published an article by society member Simon Ball titled "Further Data on the Pilbara Death Adder—a Suspected New Species," which noted that Ken Aplin was working on a formal description. "He even had a name selected," Ball told me. The species was to be *Acanthophis abditis*, from the Latin for "hidden." "It had been hidden from science, in plain view, if you like," Ball said. "And he was going to describe it as that. And we were all calling it that among us."

But years passed. Aplin's description didn't appear. "I'm not sure what held Ken up," Ball said. "He had many things on his plate, no doubt. Ken was the sort of person who had a million things on the go. But I don't know what held him up from publishing."

Brian Bush, a snake man and holder of the "Original Lizard in a Hat" trademark, was one of the people who knew about the undescribed death adder. Since the mid-1970s, he has spent most of his time in the outback in Western Australia, including the Pilbara region, where he teaches mine workers how to handle venomous snakes. When I spoke with him on the phone, he had just rolled out his swag near Port Hedland. He paused repeatedly to spit out flies. "I swallow a lot of them," he said. "I only eat twice a week because I get enough protein." He said that, to someone reasonably versed in death adders, the Pilbara death adder is obvious. The desert death adder, *Acanthophis pyrrhus*, has tillered scales and "a head like a mountain range—the scales are lumpy and bumpy and fragmented." The Pilbara death adder, by comparison, has smooth scales. Furthermore, Bush said, while desert death adders subsist mostly on lizards, Pilbara death adders favor small mammals.

One day in 1997, five years after Ken Aplin appeared in a newspaper holding the snake he planned to name *Acanthophis abditis*, Brian Bush got a call from Raymond Hoser, "out of the blue," he says. The two men knew each other from various reptile conferences. Hoser had just become editor of *Monitor*. Bush says Hoser asked him, "Hey Busho, what's this death adder from the Pilbara that Ken Aplin's working on? What's different?" Knowing what he knows now, Bush says, he would not have taken the call, much less answered that type of question. But at the time, he says, "I didn't think anything of it." Bush says he told Hoser that the new snake had smoother scales than the desert death adder, and that it had two prefrontals—the scales directly above its nostrils—instead of four.

A few months later, Bush received the latest issue of *Monitor* in the mail. It contained an article by Hoser, titled "Death Adders (Genus *Acanthophis*): An Overview, Including Descriptions of Five New Species and One Subspecies." One of the new species that he had raised from scientific oblivion into the shared Linnaean tongue was the Pilbara death adder—the snake I would much later recognize in *Snakes of Western Australia*, the book Hoser pulled down from his shelf near the end of a very busy day. He had named it after Richard Wells, and in the process forever appended the species with his own name and the year he'd named it: *Acanthophis wellsi* Hoser, 1998. The species could be distinguished from the desert death adder, Hoser wrote, by the lack of keeling on its body and head scales and by its two prefrontal scales.

"Blow me down," Bush recalls thinking. "I nearly fell out of my chair."

~

In 1999, the year after Hoser described the Pilbara death adder and four other new death adder species in *Monitor*, Ken Aplin and the South Australian Museum's Steve Donnellan published an article in *Records of the Western Australian Museum*, titled "An Extended Description of the Pilbara Death Adder." In the paper, they offered a head-to-tail physical description of the snake, from its subcaudal scales to its rostrals. They had conducted allozyme electrophoresis of the snake's liver homogenates, and used morphological and genetic statistics to compare it with other death adder species. They included a dozen photographs of the snake, a map of its geographic distribution, and a family tree showing its evolutionary relationship to other death adders. Aplin and Donnellan's description, in other words, was far more thorough than Hoser's, offering a much broader range of evidence to support the hypothesis that the Pilbara death adder is, in fact, its own species. The paper had been in peer review when Hoser's article was published, the pair wrote. But they acknowledged that Hoser's description, which included characteristics that distinguished it from other death adder species, a name, and a type specimen, appeared to "minimally satisfy" the requirements of the International Code of Zoological Nomenclature. Since his description appeared before Aplin and Donnellan's, the principle of priority meant that Hoser's name would stand.

Brian Bush realized he had inadvertently provided Hoser with the distinguishing details that Hoser had used to describe the Pilbara death adder. "My nose was out of joint," he said. "I was angry. I was angry because I was basically made a fool of." (Hoser's paper cites an email exchange with Bush as the source for those details, not a phone call; Hoser told me he doesn't remember either the call or the email, but that he had known the distinguishing details for years, and that any question he may have had for Bush was to seek corroboration, not new information.)

"We felt sorry for Ken," Simon Ball said. "Ken was probably disappointed, but he never really showed it."

Steve Donnellan, Ken Aplin's coauthor, did not respond to any of my requests for comment. Aplin died in 2019, but he made his views clear in an article he published in 1999 in *Monitor*, titled "'Amateur' Taxonomy in Australian Herpetology: Help or Hindrance?"

Amateur taxonomy remained rare in Australian herpetology, Aplin wrote—Hoser's revision of the death adders and snake man John Cann's description of four new turtle species, published the year prior, marked the first "publication of taxonomic works by private individuals in non-academic journals" since Wells and Wellington's infamous publications of the mid-1980s. Cann's descriptions, Aplin wrote, "are perfectly adequate and either contain or point to all of the information necessary for independent assessment." But Hoser's descriptions, he wrote, "are far less adequate . . . He does not present sufficient evidence to support the recognition of a further five species on top of those currently listed."

Aplin wrote that, like Wells and Wellington's papers, Hoser's work "has a distinctly 'anti-institutional' flavour, spiced as it is with frequent reference to wildlife enforcement issues, alleged corruption and tardiness/oversights on the part of 'institutional' herpetologists." Both efforts seemed to stem from a frustration with the obvious number of still-undescribed species and with the slow progress by Australia's professional herpetological taxonomists, the museum- or university-based researchers who, by Aplin's count, numbered only around twenty people at that time. "Unfortunately, as demonstrated by Hoser's attempt to 'fast-track' a revision of *Acanthophis*, there are no real shortcuts in the business of taxonomy."

Aplin ended his article with a plea for unity. "If we are going to document the true diversity of [reptiles and amphibians] in this country, we are going to have to work far more closely together. . . . 'Professional' herpetologists will need to broaden their view of 'amateurs', away from the prevalent attitude of today—that 'amateurs' can be trusted to do useful fieldwork and captive breeding studies but not much else. . . . And 'amateurs' need to abandon any tendency to view the 'professionals' as some elitist club, or worse still, as a gang of callous murderers—the antithesis of their own love for herptiles."

He concluded, in a curious augury of what was to come: "The international scientific and herp communities have already seen more than enough of Australian 'amateurs' and 'professionals' doing battle."

In the same issue of *Monitor*, Hoser offered a response. Aplin was correct in noting, he wrote, that he had not inspected the death adder specimens held in all Australian museums, that he had conducted no genetic testing of specimens, that he did not personally inspect the individual animal that he designated as the type specimen of *Acanthophis wellsi*, and that he did not consult with Aplin, whom he knew to be working on a description of the species. "The fact remains," he wrote, "that none of the above is essential for the publication of valid species descriptions." In other words, whatever gentlemanly norms he'd broken, whatever standard procedures he'd ignored, he had followed the rules inscribed in the International Code of Zoological Nomenclature, which, when it comes to applying names to new species, are the only rules that really matter.

Despite Aplin's critiques, Hoser continued, "he played the ball, not the man and I respect him for this." But others had been less kind. "Most of this 'other' criticism was beneath contempt," he wrote; much of it came from "the group I would call 'the usual suspects.' These included the bent officials, crooks and allies detailed in my books *Smuggled* and *Smuggled-2*. Being a corruption whistleblower has never made me popular amongst the corrupt, and these same people never cease to miss an opportunity to stick knives in my back."

This article seems to mark a moment of convergence, when the aggrieved, rebellious tone of Hoser's whistleblowing fully subsumed the relatively sober and dispassionate tone that had previously characterized his scientific works. He told me that the criticism of his first taxonomic work was motivating. It filled him with defiance. "My initial intent was never to name more than the death adder," he says. "If they said, 'Raymond, you know, great job naming those death adders,' that would've been the end of it. They created a monster."

CHAPTER 8

SNAKEMAN

Danger, thrills and excitement
Lie in a snake-man's job:
Milking, displaying deadly snakes
They are a brave mob.
"Where have you been today, snake-man?"
"Out in the bush," he replied;
"Caught an adder but if he bit me
I surely would have died."

—"SNAKE MAN," *SYDNEY MORNING HERALD*, OCTOBER 12, 1980;
$4 PRIZE AWARDED TO AUTHOR JIM COPLAND, AGE TEN

Our drive across Melbourne was marked by numerous navigational setbacks, which Raymond Hoser took more or less in stride. At one point, after Google Maps announced adverse traffic conditions ahead, he explained that because of his years of driving a taxi, he would be able to reduce the projected ten-minute delay to a mere two minutes, a prediction he nearly made good on. He weaved between lanes and noted the location of police officers and traffic cameras and five-gallon buckets and applauded pedestrians for their fitness and leaned out the window to woof at dogs and never paused in his explanations of search-engine optimization and the relative merits of Google and Yahoo and Bing and the workings of the international Bern and Madrid treaties and the intricacies of

American state and federal copyright law. We turned off the highway onto side streets and he slowed only a little, straightening roundabouts and rolling over curbs. Behind us, the plastic tubs rattled and bounced. We made a fast right onto a narrow residential street and Google Maps announced that our destination was on the left. Hoser swung a U that caused the Nissan Micra's tires to screech, and then we jerked to a halt opposite the house. Still talking about trademarks, he exited the car and pulled on a rumpled black polo with yellow and red lettering that read: "SNAKEBUSTERS™," "Australia's Best Reptiles™," and "Hands On Reptiles™." Then, our arms full with the plastic tubs, we walked into the party.

The birthday girl's name was Eva. She was seven years old. She wore green shorts, a black T-shirt, and a pink bucket hat. After ascertaining her name and age, Hoser picked her up and flipped her upside down and then swung her around by the arms. This is something Hoser does at nearly every birthday party he works. On YouTube, there are videos of him hoisting and turning upside down adult men, teenage girls, a Santa Claus. We set up under a blue folding gazebo. The rest of the backyard was occupied by a swing set and a weary inflatable pool-and-play structure.

"Now, kids, count of three, we're going to say 'Happy birthday, Eva!'" Hoser told the dozen-odd children who had gathered in a circle under the gazebo. "One, two, three." The children gave a lackluster cheer. "I can fart louder than that," he told them. He asked who wanted to hold a crocodile. A boy named Noah volunteered. "Kid," Hoser said to him, "what do you call a kid under the age of two?"

A toddler, Noah said.

"A lockdown baby," Hoser corrected him. During the COVID-19 pandemic, Melbourne endured a grueling series of lockdowns, which is something that he brings up frequently. "So you hold this crocodile like a lockdown baby," he said.

After passing around Joshua the crocodile, Hoser pulled out a blue-tongue lizard, which was flicking its tongue. "Who wants a lick? Come on, young child," Hoser said. "No one dies being licked by a blue-tongue lizard." He passed around the lizards, then green tree frogs, some of which were actually purple. He asked the children if anyone knew why a green tree frog might turn purple.

"Because someone farted on it," said one of the children. The child's parents explained that their family had seen Hoser's act the year before.

Hoser then received a call on one of his four phones. "Helllll-o!" he said. "How can I help you?" It was another potential client with another potential snake. He wandered off past the play structure to gather important details as children and adults continued holding and dropping green tree frogs and getting licked by blue-tongues. He wandered back. "Who has not held a frog yet?" he said, offering one. "Frog? Frog? Frog? Frog? Frog?" He offered turtles. "Mothers and dads, don't ever get turtles for pets," he said. "They're revolting. They poo, they stink, they're gross and smelly and they will stink your house up. They're the worst pets." He instructed the kids to turn the turtles on their backs. Thus upended, the turtles would immediately crane their necks ninety degrees to starboard and pop themselves right side up. This appeared to be reflexive, encoded within their very turtlehood, because they would do it no matter how many times the children flipped them over.

"Now, kids—turtles—revolting!" Hoser said as he gathered up the turtles. "Now, kids, everybody sit down and we'll do the snakes. I'll show you the dangerous ones. Now, kids, this is the best part of the party!"

~

Hoser began performing as a snake man in the early 2000s. He started by helping his friend Fred Rossignoli put on shows. At that time, Rossignoli "was the best in the business," Hoser recalls. Rossignoli, who Hoser says was also "blind as a bat," often went barefoot during shows because he sometimes accidentally stepped on the snakes and he didn't want to hurt them with heavy shoes. At the show's crescendo, Rossignoli would pile ten or twenty deadly snakes on a podium, then reach underneath them, scoop them up all together, and set them on the ground.

At some point in this apprenticeship, Rossignoli suggested that Hoser begin putting on reptile shows of his own. Hoser wasn't interested at first. He didn't think the wildlife authorities would grant him a license. But Hoser's wife encouraged him. They'd met and married in the late 1990s, and while she tolerated his herpetological interests and even dutifully visited him twice a week during his stints in jail, she was increasingly unhappy with the side effects of his career as a dauntless whistleblower. "What happened was, my wife says the book stuff is bad," he says. She knew Rossignoli, and she knew he appeared to be running a functional business and he hadn't been raided by wildlife officers. Hoser agreed to start performing. "Okay, I'll do it," he recalls saying. "But if the wildlife department shuts me down, I'll quit." Soon he was putting on shows at shopping malls and birthday parties and elementary schools. To his surprise, the wildlife officers seemed pleased. "I'd have five or six tiger snakes around my neck, and the wildlife officers were encouraging me," he says. "They'd say, 'Tomorrow do it with fifteen!'" He realized the officers expected him to be bitten, perhaps fatally.

But, unlike Rossignoli and probably every other snake performer in Australia, Hoser knew there was no chance his snakes would kill him. In a 2004 article published in *The Herptile*, he revealed the reason for his confidence: Through a surgical procedure of his own invention, he had removed his snakes' venom glands. "Noting the increase in public liability insurance problems and related occupational health and safety laws," he wrote, "the idea of using deadly snakes for displays in close quarters became problematic." He had begun his experiments on venom-gland surgery by dissecting road-killed tiger snakes he found. "While some corpses were quite decayed and smelly," he wrote, "they were still adequate for me to inspect the venom glands and test means of conducting surgery."

The technique itself was simple, although he cautioned that it should "only be undertaken on healthy well-adjusted snakes." Hoser first attempted the procedure on a juvenile tiger snake. He sedated the snake by chilling it in the refrigerator, then taped it into a specially built snake-holding jig and fixed its jaws open. He used a sterilized scalpel to remove the venom glands from the inside of its upper jaw, sutured the wound, and released the snake from the jig. Once it had healed, he let it bite a mynah bird, which suffered no ill effects, thereby proving the success of the operation.

This innovation led to what were undoubtedly some of Hoser's greatest successes as a performer, enabling him to treat audiences to a style of snake show little seen since the days when Charles Underwood, Joseph Shires, and other early Australian snake men wandered the country selling their proprietary snakebite antidotes. "These are the world's deadliest snakes," Hoser tells a group of children in a classroom in a video from the mid-2000s, as he unspools a pair of taipans from a tub. "No one's ever survived a bite from an inland taipan without antivenom. So what I'm going to do, in a world first: I won't just get bitten by one of them, I'll get bitten by them both." The mouth of a boy opposite the camera hangs open. A girl gnaws her fingertips. "Now, normally you get bitten by a taipan, you're dead in about two minutes," he says. "Now, I don't really enjoy this, but watch this. Ready and fire." He jabs the open mouth of one of the taipans into his right forearm. The children gasp and groan. As blood wells from a pair of holes, he takes the second taipan by the head and repeats the operation. "Now, in a world first, these snakes have had their venom surgically removed," he tells the children, "which means that I will still be here by the end of the night." Newspaper articles from the era report that Raymond Hoser's Snakebusters show was "unique because audience members got the chance to hold some of the deadliest snakes in the world, at no risk of being poisoned."

At Eva's birthday party, however, he did not force his taipans to bite him, nor did he allow the gathered children to touch them or even get close. Compared to the rest of his show, the venomous-snake portion—including appearances by a red-bellied black snake, a copperhead, and an eastern brown, which he tossed in the air to demonstrate its docility—felt hurried, almost cursory, and it was unclear if the children viewed these snakes as any more or less interesting than the blue-tongues. For Hoser to display his devenomized snakes at all had become something of a legal tightrope.

Many herpetologists, it turned out, were unimpressed by his groundbreaking surgical procedure. Many veterinarians and Victorian wildlife authorities were also unimpressed. Some thought the venom sacks could regrow, and that a devenomized snake might soon turn back into a venomous snake. Some thought the surgery should have been performed by a medical professional. Others thought the surgery was unethical and shouldn't have been performed at all. In 2007, the Victorian Department of Sustainability and Environment amended its rules regarding wildlife demonstrator licenses, stipulating that "venomous snakes (whether or not the individual specimen is capable of a venomous bite) must not under any circumstances be handled or touched by any person other than the holder of this license or their licensed Assistant without the prior written approval of the Secretary." It also ordered that, when handling venomous snakes, the wildlife demonstrator license holder must maintain a distance of three meters between themself and audience members, and that the wildlife demonstrator license holder may hold only one venomous snake at a time.

In 2008, Hoser wrote to the department secretary requesting permission to continue showing his devenomized snakes in his customary manner, by the armful and in close proximity to bystanders. The secretary denied these requests, so Hoser appealed his case, sending it to a tribunal hearing. A lawyer represented the department. Hoser represented himself. Anne Coghlan, deputy president of the Victorian Civil and Administrative Tribunal, presided. Following some discussion over which of Hoser's 205 items and attachments would be admissible as evidence, he presented his case. "He covered a wide range of matters," Coghlan wrote in her decision. "Much of it was repetitive and not to the point. . . . He says there has been a campaign against him to destroy his business interests."

The department's representative called a veterinarian who Hoser said had certified his devenomized snakes as completely safe and incapable of regrowing venom sacks. The veterinarian testified that he had certified no such thing, verbally or in writing. Hoser argued that it was necessary for him to hold multiple snakes so that at least one snake's head would be visible to all members of a large audience, and argued that for him to convey important taxonomic information, the audience had to be closer than three meters so they could view the snakes' "scalation and other diagnostic features." He called Fred Rossignoli to give evidence. Rossignoli testified that he personally did not allow children to hold venomous snakes, that he generally kept a three-meter distance from an audience, and that he used a whiteboard for educational purposes. Having considered this and other evidence, Coghlan denied Hoser's appeal. Soon after, he told a reporter with *The Age* that his devenomized snakes were "100% completely safe," and that he would continue fighting to display them.

After placing the devenomized snakes back in their tubs, Hoser brought out his twelve pythons. "The ones you hold are called pythons," he told the children. "Say python." He asked the children what Americans call pythons. "Constrictah," he said. He asked the children what people in New York call rubbish. "Gahbage," he said. He pulled out a python and draped it around Eva's neck, then went around the circle, draping pythons. Some were three feet long, skinny as a Wiffle Ball bat. Others were much larger. Hoser draped them without prejudice, so that hulking men sometimes wore small snakes and tiny children wore large ones.

I attended a number of children's birthday parties with Hoser, and this was always the best part. As he distributed his pythons, Eva's party immediately descended into the type of collective surreality that is usually achieved only by alcohol administered in heavy doses. There was a boy in a ball cap alone under a corner of the gazebo with three snakes coiled around him and he was gazing at the head of the albino one bobbing inches from his nose and he didn't seem to know what to do with his arms. There was a large man wearing a snake who said, "I'm not all that keen on snakes, tell you the truth." There was Eva's father Mick, who said he'd been to six or seven of Hoser's shows and preferred him to other snake men because Hoser eschewed undue formality, decorum, and structure. People draped in snakes took pictures of other people draped in snakes, and children scurried forth clutching blue-tongues and green tree frogs.

In my notes and photos and recordings from the party and in the memo I wrote that night before bed, Eva's party is a discreet event, separate and distinct. But in my memory, details from all of the children's birthday parties run together. There was the party where a woman refused to get any closer to the snakes than the front fence and another woman got no closer than the back fence, and the party where a young man with LOVE and HATE inked on his knuckles and a cigarette in his mouth could barely look at the snakes and fled the scene entirely when someone offered him one. There were parties where a kid was dressed in a mysterious orange suit like a human Cheeto and where a toddler kept trying to open the tub that contained the taipans and where a little girl holding a blue-tongue lizard went around yelling at people to look at her lockdown baby.

At this stage in the parties, Hoser was sometimes Bacchus, pouring more herpetological libations for the revelers, bestowing frogs and foisting snakes and hoisting children and calling women "lady" and men "bald man" and "large man" and children "small child" and "young child" and "kid." At other times, like when he needed to answer phone calls from prospective clients or when the taipans shit themselves badly, he wandered away from the party, allowing events to unfold without close supervision. During these times, people sometimes desired assistance in extricating themselves or their child from the grip of a constrictor, or simply wanted not to hold a blue-tongue any longer. Seeing that I had arrived with Hoser, many of them assumed me to be his assistant, and so I often found myself taking notes while draped with one or more pythons. I liked the pythons. They were dry and smooth, with the fluid mass of a weighted blanket. Some were the color of weathered ivory, some a checkered tan, some a deep brown olive. When the sun caught their scales, they had the shimmering, shifting chatoyance of a violin back. When they were relaxed, just hanging out, their bodies felt fleshy and soft. But when I tried to disentangle them from the straps of my backpack or from a child's leg, they turned into Charles Atlas's bicep.

During one of the parties, Hoser found a piece of chalk and asked the host whether he might draw on the sidewalk. He then drew several large stick figures. One was of his wife. The other two were of his daughters. The stick figures of his wife and one of his daughters were farting. The only time I heard Hoser laugh harder was a few days later, when, at his urging, I ate a live witchetty grub. He had discovered two of them under a rock, and I thought he was going to eat the other one, but he didn't. The witchetty grub was a couple inches long and squirmed when I popped it in my mouth. It was simultaneously crunchy and mushy and grainy, and it tasted, I imagine, just like bird shit. Now, surveying his chalk drawings, Hoser was similarly undone with glee. He cackled and danced and pointed them out to anybody who might have missed them. Recovering himself momentarily, he said to me that he should also draw one of his archnemesis, the German snake-venom expert Wolfgang Wüster.

~

Wüster, when I reached him by video call at his home in Wales, where he is a professor in the Bangor University school of environmental and natural sciences, seemed somewhat perplexed to find himself in the role of chief antagonist. Indeed, aside from the winsome alliteration of his name, it is not exactly clear why Hoser has come to see him as the ringleader of his numerous enemies. Wüster, who is bald and stout and wears glasses, spoke calmly and chose his words carefully and did not swear, and struck me as generally no different from any of the hundreds of other scientists I've spoken with over the years. This was in marked contrast to Hoser, whose idling speed is about 3,000 rpm, who seems unburdened by the difficulty of choosing between words, who spices his speech liberally with constructions like "bullshit," "batshit," "piece of shit," "lunatic shit," "shithead," and "shitcanning." Wüster bore both a physical and, it appeared, constitutional resemblance to Hugh Strickland, lover of rules and order, who laid the foundation for the modern codes of taxonomic nomenclature. Yet it seemed to me there was also a hint in him of the snake man, something slightly subversive in his willingness to remain in the fray with Raymond Hoser, and in the amusement he seemed to find in the whole thing.

Wüster told me he followed the usual path to herpetology, becoming interested in snakes as a boy. "I was fascinated by the fact that most people really detested snakes, including members of my family," he said. "My grandmother would basically leave the room screaming if I opened an animal book on the snake page. I guess at some point you just think, 'Hey, there's going to be something cool about these.' And it kind of just stuck with me right through life." Wüster studied zoology in college, then wrote his PhD dissertation on the taxonomy of the cobras of Asia, which turned out to be far more diverse than previously understood. "They were one species when I started," he said, "and eleven species when I finished."

He first learned of Raymond Hoser through Hoser's *Reptiles and Frogs of Australia*, which he said was "actually quite a good book." They met in person in the early 1990s, at the World Herpetological Congress in Adelaide, but didn't spend much time talking. Hoser returned to Wüster's notice in 1998. Soon after describing *Acanthophis wellsi* and other death adders, Hoser published a description of a new genus and species, which he argued had previously been confused with the king brown snake. He called it *Pailsus pailsei*, after Roy Pails, another Australian amateur herpetologist and snake breeder. Wüster was familiar with king brown snakes, and he was skeptical of Hoser's evidence. He began paying attention.

In 2000, in *Ophidia Review*, Hoser described seven new python subspecies, two new python species, and two new python genera. These included *Lenhoserus*, named after his father Len; *Katrinus*, named after his mother Katrina; *Aspidites adelynensis*, named after his eldest daughter Adelyn; and *Leiopython hoserae*, named after his wife Shireen. In 2001, in *Litteratura Serpentium*, the magazine of the European Snake Society, he described a new species of snake that he called *Pailsus rossignolii*, after Fred Rossignoli. In the next issue of *Litteratura Serpentium*, Wüster and several colleagues offered the first of what would be many published responses, in an article titled "Taxonomic Contributions in the 'Amateur' Literature: Comments on Recent Descriptions of New Genera and Species by Raymond Hoser." As Ken Aplin had done in his response to Hoser's description of the Pilbara death adder, Wüster and his colleagues dismissed the "artificial dichotomy" between amateur and professional herpetologists, whom they described as "non-institutional" and "institutional," respectively. "None of our criticisms of Hoser's work are intended to detract from his other achievements and contributions," they wrote, "nor do we wish to belittle his considerable knowledge of the Australian herpetofauna."

However: While Hoser's names were valid according to the rules of the International Code of Zoological Nomenclature, his "descriptions are much less convincing," they wrote. "Hoser almost invariably fails to provide adequate information on his species. . . . We are in effect asked to accept his species without being able to check his data." They lamented his Latin, disputed his apparent belief that "bestowing names on potentially valid taxa constitutes a service to herpetology," and questioned his ethics. While "we regard the distinction between 'amateur' and 'professional' herpetologists as largely spurious," they concluded, "Hoser's actions threaten to make the gap a reality."

Hoser responded in *Crocodilian*, the journal of the Victorian Association of Amateur Herpetologists, describing the scientific back-and-forth as "a sinister tale of vendetta, politics, pseudoscience and a serious case of outright scientific fraud." At the end of the lengthy article, which dealt with snake taxonomy, wildlife smuggling, police corruption, and the International Code of Zoological Nomenclature's nonbinding code of ethics, he concluded, in a strange inversion of what was to later occur, that the sooner Wüster and his colleagues and "their perverse and warped attitudes are banished from herpetology, the better."

Many of the entries in the works-cited section of Hoser's *Crocodilian* article were links to posts on various internet forums, including king-snake.com, where herpetologists like Paul Hackett, Bernard Frome, Pete Brammell, W. Hawke, and Jim Paull often took his side. I could not find contact information, *curricula vitae*, or birth certificates for any of these people, and Wüster and others who frequented the forum seemed to doubt that they existed at all. The kingsnake.com forum archives remain online. They are difficult to navigate and littered with broken links, providing a perspective that is, on the whole, fragmentary and impressionistic. Still, they offer a window into the early days of the feud.

Posted by WW on November 22, 2002 at 03:39:52

I have never stated that P. pailsi is definitely P. australis, I have stated that your description does not provide adequate evidence. We have been through that many times. . . . Had you provided sufficient hard data in your original paper, then you would have had fewer doubters to contend with.

Cheers,

Wolfgang

Posted by rayhoser on November 22, 2002 at 05:00:47

Well, I was obviously being too hopeful for you to admit your mistakes in the first instance and hence your baseless claims that my original description/s were/are inadequate, nomen nudems, unethical (?), or otherwise tainted is totally understandable in that you must somehow justify your fundamentally unjustifiable comments of the past. . . .

IN ANY EVENT, I BEAR NO MALICE AGAINST YOU AND ARE WISHING YOU ALL THE BEST

Raymond Hoser—Australia

Posted by WW on November 23, 2002 at 05:04:02

Please quote the passage in which you discuss the ventral count of P. rossignolii and use it to distinguish this species from P. pailsi or P. australis.

Thank you.

Cheers,

Wolfgang

Posted by rayhoser on November 23, 2002 at 15:00:48

. . . Your methods are fraudulent at best and scandalous at worst and in time you will stand condemned for this. . . .

Thus I will make this my last contribution to your mischievous and misleading thread, in which you tell bare-faced lies in relation to the original Pailsus descriptions in order to improperly tarnish my reputation. . . .

Bye for now

~

In 2002, Hoser named a new subspecies of taipan, and three new species and two new subspecies of death adder. In 2003, he named four new species of rough-scaled snake, brown snake, and python, and seven new subspecies of black snake and python, including *Chondropython viridis shireenae*, after his wife, and *Katrinus fuscus jackyae*, after his youngest daughter. In 2004, he named two new python genera and six new subspecies, including *Shireenhoserus*, also after his wife. In 2005, he named a single new subspecies. In 2006, 2007, and 2008, he named no new creatures. He says he simply didn't have time.

His business was booming. He was managing ten employees, he says, who drove cars emblazoned with the Snakebusters logo. They performed multiple shows every day at primary schools, secondary schools, childcare centers, malls, and other places of business, education, and leisure. They fielded nearly constant requests from people who wished to have snakes removed from their homes.

During this time, Hoser was also hard at work defending his various trademarks. These efforts included a lawsuit against Prospero Productions, which produced a television documentary called *The Snakebuster* starring a Western Australia snake catcher, and letters to numerous newspapers and media demanding they apologize for using the term "snakeman," "snake-man," and "snake man" in reference to other snake men. The phrase "snake man" appeared in Australian newspapers at least as early as 1827, when a writer used it in reference to a man whose "familiarities with the snake family have entailed upon him the characteristic appellation of the 'snake man.'" Over the next two centuries, newspaper writers, lawmakers, and the Australian public regularly used the term, always referring to a man who habitually handled snakes. But Hoser rejected the notion that "snakeman" is a vocational description akin to "policeman," "foreman," and "handyman." "It could be a man catching snakes," he told me, "or a man dressed as a snake."

In 2009, amid this frenzy of activity, Hoser made a remarkable innovation, which achieved for his taxonomic efforts roughly what the Gutenberg press achieved for Bibles: He founded the *Australasian Journal of Herpetology*. According to its website, the journal, published by Kotabi Publishing, a company owned by Hoser that also published several of his books, has a number of advantages over other publications, including strict procedures "to avoid undue censorship of material that may be unpleasant to vested or powerful interests," a thorough peer-review process, and simultaneous online and print publication, ensuring compliance with the International Code of Zoological Nomenclature. The website also includes detailed submission guidelines covering font size, citation style, the peer-reviewing process, and the need to clearly distinguish opinion from fact. These guidelines seem even more thorough considering that, in the seventeen years the journal has existed, Hoser is the only author it has published. He insists that every paper is peer-reviewed, although when I asked how I might get in touch with these reviewers, he said it wouldn't be possible: "I don't know you from Adam."

In form, the *Australasian Journal of Herpetology* is classically Hoserian, with colorful photos, bold fonts, and arresting headlines. The first issue consisted entirely of a paper titled "One or Two Mutations Doesn't Make a New Species," about Australian copperheads, genus *Austrelaps*. In a manner typical of scientific journals, it noted the date the paper was submitted (July 16, 2008), the date the journal accepted it (October 10, 2008), and the date of publication (January 1, 2009). Raymond Hoser was listed as the paper's sole author. Eleven of the paper's twenty-three references were to other Hoser papers. One middle page is filled with an image of Hoser holding a ball of venomous snakes above a banner that reads

**SNAKEBUSTERS—AUSTRALIA'S BEST REPTILES
IS PROUD TO BE ASSOCIATED WITH
THE FIRST ISSUE OF
AUSTRALASIAN JOURNAL OF HERPETOLOGY.**

The second issue appeared just over a month later. It also consisted of a single paper by Hoser, discussing python taxonomy and the failure of certain herpetologists and taxonomists, whom he called "truth haters" and "Hoser critics," to recognize and adopt his python taxonomy revisions. "None of their continual barrage of criticisms has had a grain of merit," he wrote. "However using their excess amounts of spare time and the near limitless resources of the internet, these [men] have managed to wage a campaign against Hoser of a scale and magnitude that is truly amazing." In the paper, Hoser also named four new python subspecies and two new python subgenera, including *Chondropython viridis adelynhoserae* and *Jackypython,* after his daughters.

In the journal's third issue, which appeared the day after the second issue, Hoser named a new genus and a new species of skink, which he named *Allengreerus ronhoseri,* after his uncle. In the fourth issue, published several days later, Hoser described nine new subspecies of snakes, including *Oxyuranus scutellatus adelynhoserae* and *Pseudonaja textilis jackyhoserae,* after his daughters. These descriptions were accompanied by a photo of the girls, ages seven and nine, wearing black polos with yellow-and-red lettering. Both were draped with taipans. In the fifth issue, published the next day, Hoser offered no new taxonomic works but did provide a lengthy discussion of snakes biting their handlers. The sixth issue of the journal appeared roughly three weeks later, in early March, and contained Hoser's descriptions of four new genera and seven new subgenera of rattlesnake, including *Hoserea,* named after his wife.

By mid-March 2009, Hoser had named roughly seventy new species, genera, tribes, and so on. Some of this knowledge was gained through a lifetime of herpetological study and was stored in Hoser's brain, he said. But he acknowledges discovering much of it stored in the papers of other scientists. "They say my science is shit," he told me. "There *is* no science to be shit."

~

In 1837, Darwin scribbled a small tree in his notebook, beside the words "I think." He later expanded on the thought in his *Origin of Species.* The broad architectural similarities and repetitions among the world's species, such as feathers, bilateral symmetry, and nipples, exist not because the Creator had a limited repertoire but because those species were related, he argued. In the same manner that a genealogical tree depicting births and marriages shows the common ancestry of a family, so could a tree depict the evolutionary history of life, every fork noting a place where one species became two. Suddenly the puzzle of taxonomy was not just of matching like with like but of unraveling the history of life on Earth.

Today, these evolutionary trees are known as "phylogenies," from the Greek *phylon*, or "race," and *geneia*, or "origin." In recent decades, advances in molecular technologies have enabled scientists not only to gauge how genetically different two or more individual creatures are from one another but also to predict, based on the estimated rates of genetic mutations, when those individual creatures last shared a common ancestor—that is, when in time each fork in the tree occurred. This, in turn, has allowed them to ground their phylogenies in time, to tie the branching of the evolutionary tree to geological or climatic events in Earth's history.

This phylogenetic perspective, focused on common ancestry, is the evolutionarily accurate one. But, as Carol Kaesuk Yoon writes in her book *Naming Nature*, it has deepened the divide between the intuitive, even instinctual, human feel of things, and the modern science of taxonomy, which has in recent years revealed all kinds of evolutionary relationships that are contrary to good sense. For instance, "reptiles" is an evolutionarily nonsensical category, because it arbitrarily excludes birds, which evolved from dinosaurs. Similarly, Yoon writes, "fish" is not a taxonomic description but an architectural one, since some fish, such as salmon, share a more recent common ancestor with humans than they do with some other fish, such as sharks.

This comparative form is a popular type of trivia in forums like Reddit, where I recently discovered a group of people regaling one another with counterintuitive facts: Spanish moss is more related to oak trees than to moss. Oaks are more related to Spanish moss than to spruce trees. Mushrooms are more related to animals than to plants. Giant kelp is more related to the organisms that cause malaria than to plants. Electric eels are more related to piranhas than to moray eels. Horses are more related to rhinoceroses than to cows or deer. Hippos are more related to whales than to horses. Bats are more related to dogs than to mice. Mice are more related to humans than to shrews. Lizards are more related to humans than to salamanders. Aardvarks are more related to manatees than to anteaters. Pangolins are more related to lions than to armadillos. Falcons are more related to sparrows than to hawks. Chard is more related to cacti than to lettuce. Lettuce is more related to sunflowers than to cabbage. Cabbage is more related to papaya than to chard.

Phylogenies can contain other types of surprises. Sometimes, when scientists perform genetic analyses of a set of creatures, they find that the DNA of those individual creatures matches neatly with previously described species. Other times, though, some of the individuals are genetically distinct from known species, different enough to perhaps be a species of their own. In the evolutionary tree, these distinct individuals appear as solitary branch tips. Often the creators of the phylogenies will mark the unnamed branch tips, signifying that the organism belongs to what may be a new species—and, often, that they intend to eventually come back and produce a formal taxonomic description. According to gentlemanly norms, such a marker is a clear signal that the creature is spoken for.

But a norm is the flimsiest of barriers. All of the required information is in the phylogeny: a potential new species, clearly pointed out; a museum or university specimen collection number of an individual organism that can be designated as the type specimen; evidence, in the form of a genetic analysis, of the species' distinctiveness. Describing creatures plucked from someone else's published phylogeny would clearly violate the International Code of Zoological Nomenclature's nonbinding code of ethics, but not the Code itself; meanwhile, the name given to that creature, according to the principle of priority, would be the one that sticks. To a person heedless of gentlemanly norms, cast out from the halls of academia, fired with a spirit of rebellion, accustomed to the taste of the king's deer, the countless phylogenies available on the internet would represent an incredible trove of creatures, only a click away, just waiting to be named.

CHAPTER 9

TAXONOMIST

The profession of science . . . attracts the half-scientists who involve themselves seriously with false principles. They are not usually inarticulate. They are zealous, foolish, and loud. They have the fixation of a great scientist but neither the knowledge nor the discrimination.

—ERIC C. ROLLS, *THEY ALL RAN WILD*

There is a moment during every party when the euphoria begins to lift and there comes a winking understanding that soon Dawn will again spread her rosy-red fingers and before long reality will return in full and words and deeds that spilled out mirthfully will be recast in shadowless midday light. So it was with Eva's birthday party. Children who had only moments earlier welcomed a serpentine embrace now squirmed uncomfortably, and they wearied of blue-tongues and green and purple tree frogs, and parents retired phones to their pockets. Finally, as the partygoers began showing signs of a collective hangover that would surely end in tears, Raymond Hoser gathered his twelve pythons, eleven lizards, six frogs, six turtles, and one crocodile and placed them back in their respective plastic tubs. Eva's parents handed him $500 cash, which he rolled up and stuffed into his sock. Then we left, flying across town to rescue a distressed citizen who had discovered what was possibly a venomous snake near her house.

During the drive, Hoser told me about the speed cameras that plague Melbourne's roads, and about the fascistic tendencies of

Australia's civil service, and about the hibernational and copulatory habits of large and small snakes, and about the comparative ease of digesting lizards and frogs and mice. He also told me about the six simple snake rules that he teaches students in his snake-handling courses. "Other people will spend days teaching different things about snakes. I don't," he said. "I only teach them six theories. But I make sure they know them backwards."

Rule 1 is safety. "Don't take unnecessary risks," he explained. "Don't get yourself bit."

Rule 2 is that snakes don't bite. "You saw what the kids were doing to the snakes," he said. "None of them were trying to bite the kids." Granted, his snakes are used to being handled, he said, and had they been wild, they surely would have bitten the children. Still, the broader point held, he said. "Snakes bust their balls not to bite."

Rule 3 is that snakes run away from people. "They want to fuck off," he said.

Rule 4 is that pain causes bites. "Have you seen pictures of Wolfgang Wüster walking around with tongs?" he asked.

I said I hadn't.

"Mark O'Shea and Wolfgang Wüster are the poster boys for snake tongs," he said. "They sell them online all the time." (Both Wüster and O'Shea—University of Wolverhampton lecturer, TV personality, taxonomist, member of the Most Excellent Order of the British Empire, alleged member of the Wüster Gang—told me they do not sell snake tongs; while snake-handling supplier Midwest Tongs does offer a "Mark O'Shea Signature Series" snake hook, designed to O'Shea's specifications, O'Shea says that he has received no payment for the company's use of his name.) Snake tongs break snakes' bones, Hoser said, which causes snakes pain, which causes them to bite. "The concept that pain causes snakes to bite is a concept I invented in 2009," he told me. "I'm sure other people thought of it, because it's almost obvious, but I'm the first to put it into writing in a peer-reviewed paper."

Rule 5 is that temperature is everything. "Too hot, they die. Too cold, they can't move," he said. "Their whole existence is avoiding the

heat or getting the heat. It's a very fine line between life and death."
For instance, he said, if he hadn't put a wet blanket over the top of the
plastic tubs in the back of the car, half of the snakes would be dead
by the time we got back to his house. Indeed, it was very hot in the
car. I asked whether the Nissan Micra had working air-conditioning.
"Nah," he said.

Rule 6 is—he paused. "Now, I can tell you the rule. Most people
don't apply it," he said. "Part of these rules isn't just knowing them
and moving on. You don't go past the rules. Everything, you apply the
rules to. Does that make sense?"

I affirmed that it made sense.

Rule 6, he said, "is that there is bullshit everywhere. Everywhere.
It doesn't come with a warning, 'Bullshit coming.' It doesn't even
come with a sign, 'This is the truth, believe it.' It normally comes so
well disguised you don't even know you're getting bullshitted to. And
the second part of that is that it's easier to tell someone bullshit than
to convince them after the fact that they've been bullshitted to and
believed it."

He continued. "So Wüster, for example . . . "

~

Through the mid-2000s, Wolfgang Wüster, the German snake-venom
expert, observed Hoser's activities from afar, offering an occasional
comment on online message boards. In 2006, he coauthored a paper
published in the journal *Toxicon*, titled "The Good, the Bad, and the
Ugly." It offered a history of Australian snake taxonomy that included
a short section on Hoser's taxonomic efforts, a number of which "were
reported to have been described with the manifest intention of scoop-
ing other researchers working on them," they wrote, "a form of behaviour generally regarded as ethically repugnant."

Then, in March 2009, Hoser's taxonomy became personal. In the
Australasian Journal of Herpetology's seventh issue, he named two new
genera of cobra: *Wellsus* and *Spracklandus*. As evidence of the genera's
distinctiveness, he cited two recent studies, one by a group of scientists

led by Christopher Kelly, and one by a group led by Wüster himself. Both studies included extensive phylogenetic trees of cobra evolutionary history that revealed distinctive, unnamed branches. Wüster says the study he led was a precursory work, and that he and his colleagues had been preparing descriptions of those unnamed branches, including a subgenus roughly corresponding to one of Hoser's new genera. "We kind of went, 'Right, he's doing this right now while we're doing this,'" Wüster says. "'What are we going to do?'"

As they pored over Hoser's descriptions, he says, they found what they thought might be an opening: Physical copies of Hoser's seventh issue of the *Australasian Journal of Herpetology* didn't seem to be lodged in five large public libraries, as required by Article 8.6 of the International Code of Zoological Nomenclature. For equally procedural reasons, Wüster and his colleagues judged the publication to also violate Articles 8.1.3, 9.7, and 9.8. No longer tethered to a validly published work, Hoser's descriptions and names would float off into oblivion, as though they had never existed.

Wüster and his colleagues decided that, based on these technicalities, they would be justified in simply overwriting Hoser's *Spracklandus*. This would turn out to be an act of great significance: It was their last attempt to stymie him using the rule of nomenclatural law and was the beginning of what swelled into an international dispute concerning, by some accounts, no less than the future course of life on Earth. But at the time, it seemed like a small gambit—all that was really at stake was a subgenus designation, nobody's idea of a big deal. So, in an article published in September 2009 in the taxonomic journal *Zootaxa*, Wüster and his colleagues argued that Hoser's descriptions were invalid, and they plastered over *Spracklandus*, describing the same group of cobras as the subgenus *Afronaja*. Then they waited.

~

Hoser didn't respond at first. He was too busy. In addition to managing employees and fending off detractors and political opponents and trademark infringers, he faced ongoing difficulties related to his devenomized snakes. As his various appeals worked their way through the legal system, he continued showing the snakes, which led to additional censures and charges. Over the next two years, he published just one issue of the *Australasian Journal of Herpetology*. It contained his whistleblowing exposé on a koala that was "allegedly injured" during the 2008 Australian bushfires and filmed drinking water offered by a fireman. In the article, titled "Sam the Scam: Sam the Koala Is an Imposter!," Hoser revealed that the video was "carefully planned and executed and not an unexpected and random act of kindness."

Then, in August 2011, he suddenly became much less busy. A video filmed the month prior surfaced. It showed Hoser and his elder daughter, age twelve, performing at the Melton Shopping Centre, before an audience that a court document later noted "largely comprised schoolchildren." I could not find the video online, but the same court document included a transcript, which I have minimally abridged for length and clarity:

"When we do a show like this what would be the worst thing is if a kid got bitten," Hoser told the audience.

If a kid got bitten at one of our shows, holy moly . . .

Going to bring up my wonderful daughter Adelyn. She has never actually been bitten by a snake in her life. She has seen me get bitten a few times. She does handle them a lot, every day of the week except on Sundays, when she goes to church. . . .

We have a permit for this, by the way. . . .

The reason we're doing this is just to show we are safe. The reason is a rival operator who is trying to undermine us is saying that our snakes have regenerated venom. Which is not possible . . .

If she is still standing at the end of this show, we know that they have not regenerated. . . .

You will see some blood. . . . Ready set go . . . In theory she should be passing out in about ten to fifteen seconds. Get your stopwatches out. The worst thing that could happen is if you killed your own kid during a snake show.

The document reported that Hoser forced a death adder and an inland taipan to bite his daughter a total of seven times. "She's been bitten many times," he later told *7News*. "She knows what it's about and she's happy. She'll take a bite for you now."

Adelyn agreed. "I play basketball and stuff and there are lots of pushes so I've had worse," she told the reporter. "It didn't really hurt."

A howl of disapprobation arose. "There's no reason for a dad to put their children through something like that," the chief executive of the Australian Childhood Foundation told the *Manningham Leader*.

"Anyone's first impression would be that this is absolutely child abuse," a psychologist said.

The deputy mayor of Manningham leapt to Hoser's defense, telling the *Leader*, "He would not use his children as guinea pigs if he thought they would be at risk." But less than a week later, she retracted her statement. "Upon further reflection, given my position as deputy mayor, I wish to express regret that a mistake was made in publicly passing judgment on a controversial issue unrelated to either council matters, or relevant State Government policy issues," she said. News of Hoser's public experiment garnered coverage around the world, appearing in newspapers from Zimbabwe to Canada.

Days after the video surfaced, Hoser woke to a knock on his door. Officers with the Department of Sustainability and Environment handcuffed him naked on the floor, he says, and, as his wife and children watched in dismay, spent nine hours hauling away animals, files, computers, and other materials. (Court documents say "several" hours.) A few days after that, the department suspended his snake handler's license. With his business shuttered, Hoser wondered if he was ruined. But there was a small upside, he realized. The forced vacation would give him time, finally, for all the scientific work he'd been meaning to do. "What I did," he says, "is I decided to write some papers."

In April 2012, he published the ninth issue of the *Australasian Journal of Herpetology*. It consisted of a single article, titled "Exposing a Fraud!," in which he finally responded to Wüster's overwriting of his *Spracklandus* with *Afronaja*. Wüster and his coauthors had been wrong about the absence of the *Australasian Journal of Herpetology* from qualifying libraries, Hoser wrote; in the lengthy appendix, he included emails purportedly showing that numerous libraries had, in fact, received copies, in compliance with the Zoological Code. He also announced that, due to greatly increased demand for his journal, which he said ran to "thousands of copies" per issue, he would begin charging ten Australian dollars per ten printed pages (plus various fees and bank charges) and would accept either cash or credit card.

Then Hoser began in earnest the work that would carry his name around the world. In his journal's tenth issue, also published in April 2012, he described two new species, one new genus, thirty-two new subtribes, thirty-three new tribes, and one new family of snake. These included Katrinina, after Hoser's mother; Maxhoserviperina, after his cousin; *Adelynhoserserpenae*, Adelynhoserserpenina, Adelynhoserserpenini, Jackyhoserina, Jackyhoserini, after his daughters; and *Morelia wellsi*, after Richard Wells. In issue eleven, published the same day, he described one new snake tribe, one new subtribe, six new subgenera, and eleven new genera, including *Hoserelapidea, Hoseraspea,* Hoseraspini, *Maxhoservipera, Richardwellsus,* and *Rattlewellsus.* In issue twelve, published a few weeks after that, he described three new subspecies, four new species, fourteen new subgenera, and twenty-one new genera, including *Adelynhoserea, Stegonotus adelynhoserae, Jackyhosernatrix, Charlespiersonserpens jackyhoserae, Maxhoserboa, Katrinahoserserpenea, Katrinahoserea, Stegonotus lenhoseri,* and *Sharonhoserea.*

On and on it went, paper after paper, name after name, among them *Hoserkukriae, Oopholis adelynhoserae, Oopholis jackyhoserae, Jackyindigoea,* Katrinahosertyphlopini, *Katrinhosertyphlops,* Lenhosertyphlopini, *Lenhosertyphlops,* Maxhoserini, *Maxhoserus,* Ronhoserini, *Ronhoserus, Wellsnatrix, Wellingtonnatrix, Crottykukrius,* Crottytyphlopini, and *Crottytyphlops,* these last three after his dog, Crotty, who was named after *Crotalus*, the name Linnaeus gave rattlesnakes in 1758. Between April and July 2012, Hoser published some 280 new taxonomic descriptions, bringing his total to more than 350.

~

In 2012, Allyson Fenwick was at the University of Central Florida finishing her dissertation, a phylogenetic study of *Bothrops*, a genus of venomous pit vipers native to the Americas, when her PhD advisor pulled her aside. "Hey, you should know about this," Fenwick says he told her. In a paper she had published a few years earlier as part of her dissertation work, she included a phylogenetic tree with several species that seemed to form a separate, undescribed genus. "So, they live in a different place," she says. "They have a different lifestyle. They were distinct. It's a species group that is probably properly a genus." She noted this potential new genus in the paper, intending to follow up later with more investigation and perhaps a formal description.

But now, without her involvement, those species had been placed within a new genus, her advisor told her. The genus had a name: *Jackyhoserea*.

"Yeah, it was annoying," Fenwick says.

This appears to have been a common experience. While a few of the taxa Hoser described, like the skink species *Allengreerus ronhoseri*, were ones he encountered in the field, many others he encountered secondhand, in the papers of other scientists. Often the exact origin of his taxonomic inspiration is hazy, but in some instances it is possible to retrace his steps. For example, Hoser lists *C. viridisadelynhoserae*'s type specimen—the specimen that defines the subspecies—as Australian Museum specimen number R129716. This same specimen appeared earlier in a 2003 study on the evolutionary history and distribution of green pythons by Lesley Rawlings and Steve Donnellan, who found that same individual animal to be the most genetically distinct of the green pythons they examined. In a phylogenetic tree in their paper, it appears as a branch tip of its own.

Dozens of Hoser's other descriptions from this period point to a 2011 study led by George Washington University taxonomist R. Alexander Pyron, titled "The Phylogeny of Advanced Snakes," which contains a sprawling evolutionary tree of more than 750 snake species, many of them arranged in genera that Hoser would later rearrange. "The results of Pyron et al. 2011, show clearly that *Aparallactus werneri* is sufficiently divergent from the type species of *Aparallactus* to warrant being placed in a separate genus," Hoser wrote in one description. "As no genus name is available for these snakes, a new genus is erected." He told me that he gleaned at least thirty new genera from Pyron's phylogeny. Pyron politely declined to discuss his perspective on the matter, writing via email that he didn't care "to spend any time looking back to the struggles of 2012." Rawlings, Donnellan, and most of the other scientists I contacted about the role their phylogenies might have played in Hoser's work did not respond to my calls or emails. I attribute this reticence both to Hoser's recent lawsuit against a group of scientists he disagrees with and to the hardening taboo among herpetologists against uttering his name in public.

Back then, though, many scientists felt freer to discuss what was happening. Word of Hoser's activities began to spread beyond the narrow world of herpetological taxonomy. The scientific journal *Nature* published an article about him. So did *Smithsonian* magazine. According to "prominent taxonomists," the *Smithsonian* reporter wrote, "Hoser isn't a prolific scientist at all. What he's really mastered is a very specific kind of scientific 'crime': taxonomic vandalism." Taxonomist Darren Naish wrote a series of articles on what he called "the Raymond Hoser problem." The podcast *Undiscovered* discussed the case of Travis Thomas, a grad student at the University of Florida who was about to name two new species of snapping turtles when he discovered he'd been scooped by Hoser, who named one of the new species *Macrochelys maxhoseri*.

"It went from being, in the grand scheme of things, a moderate annoyance," Wolfgang Wüster said, "to being something of really quite astonishing dimensions." As he saw it, Hoser was exploiting a loophole, using other people's phylogenies to fuel taxonomic descriptions that were "basically bereft of evidence, but they follow the rules of the Code." These descriptions were essentially placeholders, he said, traps set for other scientists who might later offer fuller, more convincing descriptions of the creatures in question, but be forced by the principle of priority to use Hoser's names. "It kind of gnaws away at the integrity of the entire scientific process," Wüster said. If Hoser were allowed to continue, it seemed likely that herpetological taxonomy would become "functionally uninhabitable," such a messy, contentious place that no self-respecting scientist would venture forth.

Hinrich Kaiser agreed. He is a herpetologist and taxonomist at Victor Valley College in California. Despite being a German venomous-snake expert and despite occasionally working with Wüster, Kaiser does not necessarily agree with Hoser's characterization of him as a member of the Wüster Gang, or even with the premise that such a thing as the Wüster Gang exists. Wüster, Mark O'Shea, and other purported members of the Wüster Gang also disputed its reality. Like Wüster, Kaiser watched Hoser's growing number of taxonomic descriptions in disbelief. "It was nonscience, unscientific, and I thought, 'Well, is anybody going to do anything about this or what?'" he says. In 2012, he went on sabbatical and decided that he finally had the time to do something about it himself.

In their 2009 article, Wüster and his colleagues argued that some of Hoser's taxonomic names weren't valid for procedural reasons, including that he hadn't properly filed issues of the *Australasian Journal of Herpetology* in public libraries. But such a tactic could work only once. Hoser wouldn't make the same mistake again. Kaiser and Wüster thought they were justified in suggesting a more drastic route. While the rules of nomenclature are important, Kaiser said, they are ultimately just part of a filing system, meant to serve the science of taxonomy. Now, in Hoser's hands, the filing system was threatening to capsize the science. Kaiser wrote notes to taxonomists around the world, asking them to become coauthors to his and Wüster's dramatic proposal.

"We were at a crossroads," Wüster says. "If we have a choice, where we're either loyal to the Code by following it to the letter, or loyal to the science, then I'm afraid we're going to stick with the science."

In issue fifteen of the *Australasian Journal of Herpetology*, published in July 2012, Hoser wrote that he'd caught wind of his enemies' scheme. Hal Cogger, the dean of Australian herpetology, had forwarded him an email from a mysterious address. Through a "brief forensic analysis of the electronic trail," Hoser discovered that the email had originated with Kaiser, but it was clear to him that its real author was his nemesis Wolfgang Wüster.

The email contained a draft of a paper that would later be published under the title "Best Practices: In the 21st Century, Taxonomic Decisions in Herpetology Are Acceptable Only When Supported by a Body of Evidence and Published via Peer-Review." This paper took the measured tone typical of scientific publications, but it made a stunningly direct attack. In table 1, it listed every creature Hoser had named since January 1, 2000 (along with a much smaller number of creatures named by Richard Wells). "To defend herpetological taxonomy against unscientific incursions," the authors wrote, "we propose that the herpetological community, including authors, reviewers, editors, users of taxon names in applications, and other interested parties, set aside and *strictly avoid the use of the taxon names listed in Table 1*." Instead, they continued, people should use a set of recommended replacement names, also provided in table 1.

What they were suggesting was in clear violation of the rule of nomenclatural law and the principle of priority: The whole world should ignore Hoser's names. It was vigilante justice.

~

In a strange way, Kaiser's petition affirmed what Hoser had been arguing most of his life: The authorities—in this case the supposedly rules-and-order-obsessed professional taxonomists—were the ones acting in iniquitous, unjust ways. Hoser, the whistleblower, bandit-hero, defender of amateur reptile keepers and unnamed creatures alike, had nomenclatural law on his side.

In December 2013, he appealed for help from the high court, the International Commission on Zoological Nomenclature. Rather than respond directly to the boycott effort, he addressed Wüster's earlier effort to invalidate his cobra genus, *Spracklandus*. In the appeal, designated case 3601, he asked the commission to verify that *Spracklandus* superseded the Wüster Gang's subgenus *Afronaja*. He also asked the commission to confirm that the *Australasian Journal of Herpetology* fulfilled the requirements of the Code of Zoological Nomenclature, and that the several hundred taxonomic descriptions published in its pages were therefore nomenclaturally valid. Hoser's request was offered not only in his defense, he wrote, but in defense of all other taxonomists, professional and amateur.

At risk was the system of nomenclatural rules, and therefore the centuries-old scientific field of biological taxonomy itself, and therefore conservation efforts that depended on accurate taxonomy, and therefore the continued existence of countless species, and from there, you might justifiably argue, the future path of life on Earth. As Hoser saw it, the boycott effort against him was taxonomic vandalism on the grandest possible scale.

CHAPTER 10

DAMNATIO MEMORIAE

Has your tutor let the story of Robin Hood get into your hands? Such careless rascals ought to be sent to the galleys. And has it heated your childish fancy, and infected you with the mania of becoming a hero? Are you thirsting for honor and fame? Would you buy immortality by deeds of incendiarism?

—FRIEDRICH SCHILLER, *THE ROBBERS*

For there was a man, who deliberately set fire to the temple of Diana at Ephesus, so that by the destruction of that lovely building, his name might be known to the whole world. This madness was confessed by him when he was put on the rack.... The Ephesians took care, by a decree, to abolish the memory of this worst of men.

—VALERIUS MAXIMUS

Hoser was in the middle of telling me about Kaiser's and O'Shea's and especially Wüster's never-ending perfidies and about the corruption of the judges who had overseen his various court cases and about the nigh impossibility of venomous snakes delivering nonvenomous bites—which certain so-called venomous-snake experts described as "dry bites," which in reality were "rare as rocking-horse shit," he said—when one of his four phones rang.

"Helllll-o!" he said. "How can I help you?" It was another potential client with another potential snake in their house. He had them send a picture of the snake, which he instantly recognized as a male copperhead, *Austrelaps superbus*, then he took down their address and told them he'd be about an hour.

We were on back roads now. We had suffered several additional minor navigational setbacks, but Hoser made up the time by driving so fast I could feel my spleen jumping inside me. "Everybody wants to be the number one," he told me. "And if they're not the number one, they're number two, which isn't poo. And in this country, everybody wants to be the Snakeman. But there's only one. Everyone knows me."

The phone rang again. Hoser groaned. "Helllll-o!" he said. "How can I help you?" It was the woman who had called during Eva's party, whose house we were now driving toward, checking to see if he was still coming. "We're literally one minute away from your house," he told her, and ten minutes later, we were there.

A long driveway led to a cluster of buildings interspersed with lawns, a pool, a tennis court, and an apple orchard. As Hoser parked the Nissan Micra, a flowy woman of roughly his age came out to meet us. Hoser realized he'd been there before. The woman's husband owned a snowboard shop, he remembered. He is an avid skier. On YouTube, there are numerous videos of his skiing adventures, including "Canyon Run at Mount Hotham with a Ski Crash Involving the Snakeman Raymond Hoser" and "Fastest Skier on Mountain—Fastest Snake Catcher—Snakeman Conquers Whistler BC" and "There Is Only One Snakeman in the USA, Canada and Australia."

Hoser grabbed a spare plastic tub from his car, along with a window-washing squeegee. The squeegee is useful for pinning snakes to the ground, and unlike a metal tong, its spongy edge does not risk inciting rule 4. He followed the snowboarder's wife, who is a cosmetologist, to where she'd last seen the snake, in the grass by the side of the house. The spot was in full sun now. In accordance with rule 5, Hoser surmised that the snake had retreated to the shade. The most obvious place was under the house, which sat about eighteen inches off the ground. Hoser asked the cosmetologist if she'd seen the snake go under the house.

"When I saw him, I walked away, and I don't know what he did," she said.

"All right, it could be a girl snake," he pointed out as he stomped into the deep grass at the edge of the building.

"Hey, why are you standing in there like that?" she asked. "What if it comes and bites your legs?"

"What's rule number 2?" he asked me.

I paused. I had the rules, but I did not have them backwards.

"Come on, rule number 2!" he said. "There's only six rules!"

They want to run away, I offered.

"No, you dingbat, that's rule 3," he said. "Rule 2, rule 2."

"They run away," the cosmetologist suggested.

Finally it came to me: Snakes don't bite.

The cosmetologist looked dubious. "Well, then, how do people die from snakebite?"

"Because they're fuckwits," Hoser explained.

The cosmetologist and I followed Hoser as he stalked around the perimeter of the house, then around the apple orchard, and then through the garage, where he noticed a stack of animal books. He flipped through one of them but didn't find any of the snakes he'd named. He slammed it shut. We were joined in the search sometimes by the cosmetologist's teenage niece and her friend, sometimes by the cosmetologist's dogs. After a few minutes, Hoser concluded that there was no hope of finding the snake unless he crawled under the house, which he didn't want to do, per rule 1. He told the cosmetologist to keep an eye out for the snake and to call if she spotted it. Anyway, it almost certainly was not the only snake around. The property was ideal snake habitat. "The snakes are probably thinking, 'This is fantastic,'" he said. "I'd live here if I was a snake."

The better solution, he suggested, was that she enroll her dogs in his snake-avoidance course, guaranteed to give them severe ophidiophobia. "The dog sees a snake, they'll say, 'Fuck that,' and run in the other direction," he said. He works with a professional dog trainer who administers electric shocks to the dog-in-training when it tries to approach one of the devenomized snakes provided by Hoser. Once when I called him on one of his four phones he told me he was in the middle of a snake-avoidance course and was busy hiding in a bush. Now, invoking rule 6, he warned the cosmetologist that there were other people who offered canine snake-avoidance courses that were so inferior to his that they imperiled the dogs involved, but these people had messed with his search-engine optimization to such a degree that their results popped up first, not his. "Snake avoidance—we have the trademark," he said.

He led the cosmetologist to the Nissan Micra and began pulling out tubs of snakes. He showed her a brown snake, then a tiger snake. "Why are these snakes in your car?" she asked as he disentangled his taipans.

"Children's party," he said. "Reptile parties. That's us."

"Ah," she said.

"We did a little trip out to the western suburbs and they put up with all my jokes," he continued. "They actually love me because they've had me before. Many times, actually, many times they've had me, and they knew me. They love us."

The cosmetologist was eyeing the taipans. "So, when they're nonvenomous," she said, "are they permanently nonvenomous?"

~

In March 2012, months after causing his daughter to be publicly bitten seven times by two of his devenomized snakes and just before the beginning of his most incredible taxonomic spree yet, Hoser appeared before the Victorian Civil and Administrative Tribunal to request that it set aside the Department of Sustainability and Environment's decision to suspend and cancel his licenses to catch and display wildlife. The proceedings centered on several technical points, including whether devenomized snakes could regenerate their venom glands, and on the nature of pits, barriers, and rope lines to prevent audience members from interacting directly with snakes. "The Applicant was a difficult witness," the judge wrote in her decision. "Despite the valiant efforts of his experienced Counsel to keep him within the bounds of relevancy, he would frequently digress into irrelevant material and often quite scandalous allegations." She upheld the department's decision to suspend Hoser's licenses. Hoser appealed the decision, and the matter was referred to the Victorian Court of Appeal.

During this appeal, he argued to the three presiding judges that the tribunal's consideration of his personal credibility and of whether he was a "fit and proper person" to hold a commercial demonstration license was improper; that the tribunal had conflated the concept of "pit" and "barrier," at least where snake shows were concerned; and that the tribunal had failed to consider the "negative public safety implications" of canceling Hoser's license. The judges agreed, writing that the tribunal's discussion of whether devenomized snakes might regenerate their venom glands "rarely rose above the hypothetical," and noting that "the Macquarie Dictionary defines a 'pit' as, relevantly, 'a hole or cavity in the ground.'" The court set aside the tribunal's decision, ruling that the Department of Sustainability and Environment's suspension of Hoser's licenses had been in error.

The licenses in question, however, had already expired. Hoser applied for them again. The department again denied his application, writing that he was not a "fit and proper person" to hold such licenses. He appealed, and the matter again went to the state tribunal. Hoser represented himself during seven days of hearings. In his decision, the judge took a notably weary tone, writing that Hoser was "argumentative and bombastic" but these were "not necessarily attributes which disqualify him from holding licenses." He overturned the department's decision. "I am satisfied that the applicant possesses significant ability in the husbandry and handling of snakes," he wrote. He noted that Hoser "displayed an extensive knowledge of reptiles. . . . He has conducted extensive work on the taxonomy of snakes. He has come in for criticism from academics in relation to his methods of taxonomy. This is part and parcel of work in such an area."

~

After saying goodbye to the cosmetologist, who invited Hoser and his wife to come for dinner sometime, we again sped across town in the Nissan Micra. As we drove, Hoser critiqued the snake-handling skills of the late Steve Irwin, world-famous star of *The Crocodile Hunter*—"Crikey, crikey, crikey! Smash, smash, smash!"—and told me about how his enemies, in their relentless campaign against his devenomized snakes and their simultaneous promotion of snake tongs, had spread erroneous notions far and wide, in accordance with rule 6, leading countless people to cause snakes pain, in violation of rule 4, causing snakes to bite, in violation of rule 2. Hundreds, perhaps thousands, of people had died as a result. "This isn't just in Australia," he said. "This is worldwide, because my profile is worldwide."

We turned off the highway onto residential streets, careening into a neighborhood of densely set houses. When we found the address, Hoser parked, jumped out, grabbed his tub and squeegee, and ran up the driveway. He knocked on the door, hesitated briefly, then opened it and went in. We were met in the hallway by a couple and their adult daughter. The man led Hoser through the house to the patio. The snake was ensconced below the barbecue, he said. Hoser made a few exploratory pokes with his squeegee, then set it aside. He reached under the barbecue with his bare hand and emerged with a copperhead. He held it by the tail and jiggled it from side to side like a spaghetti noodle. Another snake catcher later told me that the jiggling technique, meant to discombobulate the snake, is unique to Hoser, as are the window-washing squeegee and plastic tub. Hoser dumped the snake in the tub and slammed the lid shut. We had arrived at the house perhaps fifteen seconds earlier.

Hoser asked for a glass of water, then we all stood in the living room while he and the man discussed the snakes of South Africa, where the family was from. "Black mambas are pleasant," Hoser said. "Fuck, they don't have an aggressive bone in their body. My favorite snake in Africa, it would be the black mamba."

"Okay," the man said.

"Black mambas, they're just awesome, they're just awesome," Hoser said.

"But they bite," the man said.

"No, they don't!" Hoser said.

"Well, that's why they kill so many people," the man said.

"No, they don't kill!" Hoser said.

Hoser walked the family through how to leave a favorable review for Snakebusters on Google, then offered to show them his reptiles. We all went out to the Nissan Micra. Hoser brought out Joshua. He placed her on his head so they could take a photo. Joshua peed. Crocodile urine ran down the side of his face and neck and onto his shirt. He hardly seemed to notice. Then he passed her around. "Hold her like a baby," he instructed. "Like a lockdown baby."

Hoser collected his $200 snake-catching fee, and we bid the family farewell. As they walked up the driveway, Hoser hauled the Nissan Micra onto the curb opposite their house with a jolt that threatened to dislodge another of its hubcaps, then drove down the sidewalk with the passenger-side wheels on the grassy median. We immediately encountered a low-slung tree, whose branches screeched as they dragged along the roof of the car, then we made the turn and rocketed back down the street the way we'd come. At the first intersection, Hoser pulled to the side of the road. We were near one of his mice guys. He brought up a text message on one of his four phones and showed it to me. "Do you need any mice mate? Cheers," it said. He called the number. It rang and went to voicemail. Hoser groaned. "When you have snakes, it's like being a drug addict," he said.

We drove out of the neighborhood and back onto the highway. Hoser was telling me about how he'd successfully petitioned YouTube to delete eight hundred channels belonging to one of his trademark-infringing enemies and about the heated competition among snake men to secure gigs at elementary schools and kindergartens, when suddenly he cut across several lanes of traffic and skidded to a halt on the left side of the highway. He turned on his hazard lights, put the Nissan Micra in reverse, and, while explaining the workings of internet advertising, drove backwards down the side of the highway at full speed for perhaps a quarter mile. He stopped, opened the door, plucked a small black fedora off the road, and crammed it onto his head. "Hat," he said. "Right, okay, so. Because I'm bald I need hats."

We drove up the highway a few more miles, when Hoser again suddenly pulled to the side of the road. This time he drove down a small gravel path lined with high scrub and piles of trash. "See all this shit here?" he said. "It's a great place for a snake." He parked and grabbed the tub with the copperhead from the South Africans' home. I followed him down a narrow trail through the brush. He stopped where the path opened to a wide field. "Now watch this snake," he said. "It'll fly." Holding the tub out in front of him, he undid the latches on the lid. The tub, now lidless, dropped to the ground. The snake flickered like candlelight, up and out of the tub, into the grass and away.

~

In 2014, while Hoser fought to restore his snake handler's permit, the *Bulletin of Zoological Nomenclature* began publishing letters regarding case 3601, his plea that the International Commission on Zoological Nomenclature affirm both the validity of his cobra genus *Spracklandus* over the Wüster Gang's subgenus *Afronaja* and the overall Code compliance of the *Australasian Journal of Herpetology*.

Many of these letters took the legalistic, detailed approach that seems to come naturally to taxonomists, who are accustomed to counting subcaudal scales and comparing the shape of partly inflated hemipenes. One taxonomist wrote that any descriptions published in the journal were clearly in violation of Zoological Code Articles 8.1.1.1 and 8.6. Another argued that the journal was likely in violation of Article 9.7. Hinrich Kaiser, lead author of the boycott effort against Hoser, wrote that, based on the slight imperfections in the copy of issue seven at the Australian National Library, it appeared that the document was produced on a desktop printer, not as part of a one-hundred-copy press run, as Hoser had claimed, which meant the journal was out of compliance with Article 8.1.3. Furthermore, the position of a staple in the journal suggested hand-stapling, Kaiser wrote. "If this document were to be handled frequently, even if only to open it for reading, there are potential problems with the fastening and the paper itself."

Several herpetologists, including some who happened to be current or former Snakebusters employees, wrote letters in Hoser's defense. So did Ross Wellington of Wells-and-Wellington fame. "From my observations of the hard copies of AJH including Issue 7 and, irrespective of the contents," he wrote, "it is most definitely published and the standard of print medium is equal to or better than most other similar journals currently available."

Other commenters took a broader view. They included Hoser himself, who, in concluding a lengthy refutation of his enemies' myriad errors and misrepresentations, wrote: "It is in the interests of long-term nomenclatural stability that the Commission act decisively. . . . Failure to do so will destabilize taxonomy and nomenclature. The issue is not 'Hoser' but the stability of the Code."

In the same issue of the *Bulletin*, some seventy scientists from around the world, including Wüster, O'Shea, Kaiser, Robert Sprackland (namesake of Hoser's *Spracklandus*, the genus at the center of the case), and E. O. Wilson, the late eminent ecologist, published their own comment, offering a similarly sweeping assessment. "We believe Case 3601 represents a tipping point," they wrote. "The self-produced works by Raymond Hoser under the title of *Australasian Journal of Herpetology*, and the proliferation of names therein, are so contentious that they destabilize and cause confusion in the entire system of nomenclature, and undermine the scientific reputation and credibility of the discipline of taxonomy."

The one thing everyone agreed on was the stakes: If the commission didn't rule in their favor, there would surely follow a descent into nomenclatural bedlam, taxonomic instability, and a weakening of conservation, medicine, biology, and even science itself.

Given these stakes, the International Commission on Zoological Nomenclature took what seemed like an exceedingly long time in deciding. There were multiple reasons for the delay, says Neal Evenhuis, the aforementioned Bishop Museum entomologist and describer of *Carmenelectra shechisme*, *Pieza pi*, and other insects. He has been a member of the commission since 2015 and was the editor of the *Bulletin of Zoological Nomenclature* in 2013, when Hoser submitted his case. The *Bulletin* editor is the first to field nomenclatural cases submitted to the commission, Evenhuis explains, screening spurious claims and helping applicants hone their submissions. At any given moment, he was juggling dozens of submissions. After the *Bulletin* publishes a case, it must publish comments. Case 3601 drew an unusual quantity. According to its bylaws, the commission must then wait for a year after the last comment is published before ruling on the case. In the same period, the commission moved its headquarters from London to Singapore, causing further delay.

But Evenhuis says there was also a certain reluctance on the part of the commission to rule on the case. In their back-and-forth, Hoser and his opponents tended toward grandiosity and personal attacks, he said. "It was very weird. We didn't want to send the wrong message, as the commission didn't condone what he's doing, but we didn't want to play favorites, that we're either on his side or on the other side. We have to be an adjudicating body and be independent and arbitrary. We can take a side on a case, but not on a person."

Finally, in 2021, the commission gave its opinion. It declined to confirm that *Spracklandus* was an "available" (that is, nomenclaturally valid) genus name and that issues of the *Australasian Journal of Herpetology* were "published works in the sense of Article 8.1," as Hoser had requested. But it also declined to say taxonomists could ignore the contents of his journal, or to cast *Spracklandus* into the "Official Index of Rejected and Invalid Works in Zoology." "The Commission is reluctant to suppress, in an indiscriminate way, a large part of the work of an active zoologist," it wrote in its opinion. "The Commission operates within the strict confines of nomenclature, and judgements based on the quality of taxonomy or on ethical principles remain beyond the mandate of the Commission." Faced with a case that all parties agreed was of international import, with grave and lasting consequences for the entire enterprise of biological taxonomy, the commission delivered what was essentially a nonruling.

Hoser took it as an unqualified win. "Case with global ramifications is hailed as a major victory in favour of wildlife conservation and in the battle against scientific fraud and taxonomic vandalism," he wrote in the headline of a press release published on his website. He continued: "Better known as The Snakeman, Raymond Hoser has blue ribbon scientific credentials, having been at the forefront of wildlife research and discovery for more than 50 years. The case arose when a renegade gang of thieves . . . and a rogue university lecturer in Wales, Wolfgang Wüster hatched a plot to override the 200 year set of rules governing scientific research, discovery and naming organisms. . . . The plan, first hatched in 2009 was simply to steal works from other scientists, rename species and then claim discovery of those species. It was simply an act of personal self-gratification on a grand scale."

~

The eight years since Hoser submitted his appeal to the International Commission on Zoological Nomenclature had been his most taxonomically productive yet. Between 2013 and 2021, he described Acrantophiidae, Acrantophiini, *Acrantophis sloppi*, *Candoia aspera iansimpsoni*, Candoidiini, Candoiidae, Boigaiini, *Wellsserpens*, *Slopboiga*, *Crottyhydrophis*, *Wellspython*, *Broghammerus reticulatus dalegibbonsi*, *Broghammerus reticulatus euanedwardsi*, *Broghammerus reticulatus haydnmacphiei*, *Broghammerus reticulatus neilsonnemani*, *Broghammerus reticulatus patrickcouperi*, *Broghammerus reticulatus stuartbigmorei*, *Adelynhoserserpenae wellsi*, *Anomochilus marleneswileae*, *Macgoldrichea*, *Cylindrophis wilsoni*, *Ernieswileus*, *Cylindrophis*, *Motteramus*, *Manserpens*, *Furina ornata todd-*

pattersoni, Suewitttyplops, Notopseudonaja modesta wellsi, Pailsus hoserae, Adelynhosertyphlops, Bennetttyphlops, Buckleytyphlops, Jackyhosertyphlops, Libertadictus adelynhoserae, Libertadictus cliffrosswellingtoni, Kerrtyphlops, Mantyphlops, Pattersontyphlops, Robinwitttyphlops, Libertadictus jacky-hoserae, Sheatyphlops, Silvatyphlops, Slopptyphlops, Libertadictus rich-ardwellsi, Libertadictus sloppi, Ackytyphlops, Macrochelyiini, *Macrochelys maxhoseri, Macrochelys temmincki muscati,* Ahaetulliini, Charlespierson-serpeniidae, Charlespiersonserpeniinae, Charlespiersonserpenini, Micrelapiidae, Micrelapiinae, Oxyrhabdiumiidae, Psammodynasti-idae, Psammodynastiinae, Swileserpeniidae, Thermophiidae, Oxy-rhabdiumiinae, Swileserpeniinae, Thermophiinae, Adelynhoser-boaiina, Trachyboaiini, Tropidophiinina, Tropidophiinini, *Eseraboa, Ungaliophis panamensis lovelinayi, Adelynhoserboa, Merceicaboa, Patterson-boa, Jackyhoserboa, Robertbullboa, Rodwellboa, Wellingtonboa, Tonysilvaboa, Wellsboa, Wittboa, Benmooreus, Paulstokesus, Swilea, Mattborgus, Chris-newmanus,* Brachyophidiini, *Crottyserpens,* Melanophidiini, *Jealousser-pens,* Plectruriini, *Ackyserpens, Crottyserpens,* Oxyserpeniini, *Oxyserpens,* Rhinophiini, Adelynhoserserpenina, Adelynhoserserpenini, Bothro-cophiina, Bothropina, Bothropoidina, Calloselasmiini, Cerrophodio-nina, *Charlespiersonserpens jackyhoserae,* Crotalina, *Gerrhopilus carolinehoserae,* Hulimkini, Jackyhoserina, Jackyhoserini, Maxhoservi-perina, Montiviperina, Piersonina, Porthidiumina, Rhinocerophiina, *Klosevipera, Bitis caudalis kajerikbulliardi, Bitis caudalis swilae, Kuekus, Bitis brianwallacei, Bitis lourenceklosei, Bitis matteoae, Bitis oflahertyae, Bitis pintaudii, Bitis tomcottoni, Bitis funki, Bitis hoserae, Bitis wellingtoni, Bitis wellsi, Causus perkinsi, Bothrops lenhoseri, Bothrops mexicoiensis, Bothrops mexicoiensis maccartneyi, Pughvipera,* Calloselasmiinae, *Lowryvipera, Conantvipera, Cottonvipera, Borneovipera, Crottyvipera, Cummingvipera, Katrinahoservipera, Ninivipera, Davievipera, Blackleyvipera, Ryukyuvipera, Sloppvipera, Swilevipera,* Tropidolaemusiinae, *Simpsonvipera, Yunnanvi-pera, Boulengerina adelynhoserae, Boulengerina jackyhoserae, Hemachatus haemachatus macconchei, Benjaminswileus, Lowryus,* Georgekonstandi-nouiini, *Georgekonstandinouous, Slatteryaspus, Wellingtonaspus, Chamael-ycus euanedwardsi, Hapsidophrys daranini, Hapsidophrys pintaudii,*

Lycophidion woolfi, Chrismaxwelliini, *Chrismaxwellus, Carstensus, Drewwilliamsus, Shanekingus, Avonlovellus, Paulelliotus, Liopeltis tricolor borneoiensis, Liopeltis tricolor brummeri, Liopeltis tricolor philippinesiensis,* Rossnolaniini, *Rossnolanus, Brucegowus, Dannycoleus, Shaneblackus, Staszewskius, Drysdalia mastersii andrewlowry, Drysdalia mastersii robwatsoni, Hawkeswoodelapidus,* Bennettsaurini, *Bennettsaurus, Lucysaurea, Hulimkacordylus,* Cottonsaurini, *Cottonsaurus, Edwardssaurus, Macgoldrichsaurus, Vrljicsaurus, Funkisaurus,* Funkisaurusiini, Gerrhosauriina, Gerrhosauriini, *Hallabysaurus, Hawkeswoodsaurus, Nussbaumsaurus, Raselimananasaurus, Raxworthysaurus,* Karusasaurini, Namazonurini, *Atikaea, Slatterysaurus, Ninsaurus,* Platysaurini, Swilesauriina, *Swilesaurus,* Tetradactylusiini, Tracheloptychina, *Wellingtonsaurus, Wellssaurus, Lukefabasaurus,* Woolfsaurini, *Woolfsaurus,* Zonosaurina, Zonosaurini, *Adelynhosersaur,* Adelynhosersaurini, Amphibolurini, Ctenophorini, Hypsilurini, Intellagamini, *Jackyhosersaur,* Molochini, Physignathini, Helodermini, *Maxhosersaurus, Empugusia salvator woolfi,* Empugusiini, *Euprepiosaurus indicus wellingtoni, Euprepiosaurus indicus wellsi, Odatria honlami, Odatria hoserae, Odatria tristis nini, Oxysaurus,* Polydaedaliini, *Shireenhosersaurea,* Shireenhosersauriini, Varaniini, *Aquativaranus, Arborhabitatiosaurus, Honlamus, Varanus mitchelli hawkeswoodi, Kimberleyvaranus, Parvavaranus, Pilbaravaranus, Worrellisaurus storri makhani, Dasypeltis saeizadi, Funkiacrochordus, Acrochordus malayensis, Acrochordus mahakamiensis,* Acrochordinini, Funkiacrochordidini, *Vetusacrochordus, Honlamagama, Mantheyagama, Denzeragama, Daraninagama, Doongagama, Eksteinagama, Maxhoseragama, Rubercalotes, Ghatscalotes, Laccadivecalotes, Ceyloncalotes, Tamilnaducalotes, Crottyagama, Freudcalotes, Khasicalotes, Amboncalotes, Skrijelus, Notacalotes, Jamesschulteus, Pethiyagodaus, Manamendraarachchius, Ferebronchocela, Bronchocela harradineus, Olorenshawagama, Proboscisagama, Mindatagama, Macguiredraco, Philippinedraco, Engannodraco, Somniadraco, Spottydraco,* Dracoiini, Maxhoseragamiini, Maxhoseragamiina, Sitanaiina, Acanthosauriina, Saleaiina, Crottyagamiini, Daraninagamaiini, Pethiyagodaiini, Pethiyagodaiina, Doongagamaiina, Japaluraiini, Lophocalotesiini, Phoxophryiini, Mantheyiini, Dendragamaiini, *Ferepelochelys, Pelochelys clivepalmeri,*

Pelochelys telstraorum, Aspidomorphini, *Aspidomorphus muelleri tamisi, Aspidomorphus interruptus macdowelli, Aspidomorphus keneficki, Aspidomorphus coggeri, Walmsleyus, Walmsleyus anstisae, Pilgerus, Pilgerus assangei, Pilgerus macki, Pilgerus nardellai, Pilgerus mooreae, Smythkukri hunneangorum, Cacophis scanloni, Cacophis sheai, Macconchieus, Lukefabaus, Charlespiersonserpens charlespiersoni, Chrysopelea ornata caerulea, Chrysopelea sinhaleya ghatsiensis, Chrysopelea paradisi borniensis, Chrysopelea paradisi johorensis, Chrysopelea paradisi tepedeleni, Broghammerus reticulatus mandella, Snowdonus, Acanthophis wellsi hoserae, Acanthophis pyrrhus maryani, Acanthophis pyrrhus moorei, Acanthophis antarcticus granti, Acanthophis groenveldi mumpini, Acanthophis macgregori, Acanthophis yuwoni, Paralycodon, Kotabilycodon, Myanmarelfakhari, Dannyelfakharikukri, Sinoelfakharikukri, Apollopierson, Snakebustersus, Mindanaosnakebustersus,* Snakebustersusini, *Korniliostyphlops, Supremeuromastyx, Dallysaurus, Newmansaurus, Dicksmithsaurus, Stokessaurus, Mooresaurus, Borgsaurus,* Uromastyxiini, Borgsauriini, *Supremechelys, Chelodina expansa brisbaneensis, Chelodina duboisi, Tiliqua gigas glennsheai, Tiliqua gigas grantturneri, Hydrosaurus alburyi, Odatria glebopalma funki, Odatria glebopalma maderi, Eastmansaurus, Rossnolansaurus, Kendslider, Garyallensaurus, Conningsaurus, Dannybrownsaurus, Toscanosaurus, Artusbrevis, Masonnicolasaurus, Laurielevysaurus,* Diploglossiini, Toscanosauriini, *Binghamsaurus, Smythsaurus, Richardsonsaurus,* Ophisauriini, Anguiini, *Pitmansaurus,* Pitmansauriini, Gerrhonotiini, *Lindholtsaurus, Rayplattsaurus, Rentonsaurus, Elliottsaurea, Lanisaurea, Assangesaurus, Snowdonsaurus, Namibtyphlosaurus, Marleneswilea, Kalahariacontias, Typhlosauriina, Culexlineatascincus, Ophiomorus macconchiei, Starkeyscincus, Pelleyus, Pelleyus pelleyi, Pelleyus raithmai punjabensis, Starkeyscinciini, Starkeyscinciina, Culexlineatascinciina, Parabrachymeles, Parabrachymeliini, Mexicoscincus, Mississippiscincus, Floridascincus, Californiascincus, Bermudascincus, Funkiskinkus, Marmolejoscincus, Forestaescincea, Veracruzscincus, Funkiskinkus funki, Funkiskinkus dixoni, Asiascincus, Japanscincus, Ryukyuscincus, Sichuanscincus, Adelynhoserscincea, Sinoscinkus, Jackyhoserscincea* Adelynhoserscinciini, Adelynhoserscinciina, Asiascinciina, Funkiskinkiina, *Moroccoscincus,* Eumeciini, Janetaescinciini, *Efossoka-*

lahari, Ebolaseps, Notascelotes, Parascelotes, Brygooscincus, Madascincus nosymangabeensis, Sloppyscincus, Commendatscincus, Degenerescincus, Clarascincus, Comoroscincus, Crottyscincus, Oxyscincus, Rubercollumus, Roseacaudatus, Rubercaudatus, Rubercaudatus edwardsi, Cummingscincea, Cummingscincea cummingae, Cummingscincea demiperkinsae, Gracilescincus, Leucolabialus, Gongylomorphiini, Gongylomorphiina, Chalcidiina, Sloppyscinciini, Sloppyscinciina, Paracontiina, Sirenoscinciina, Hakariina, Scelotiina, Feyliniina, Nessiini, *Aprasia parapulchella gibbonsi, Aprasia inaurita rentoni, Laemanctus viridis, Laemanctus deborrei tuxtlasensis, Brunaviridisaurus, Duboislacerta, Greerlacerta,* Atlantolacertiini, *Liasis papuana sharonhoserae, Liasis papuana cyrilhoseri, Pantherosaurus maxhoseri, Worrellisaurus primordius dalyi, Pantherosaurus giganteus queenslandensis, Pantherosaurus giganteus bulliardi, Demansia johnscanloni, Demansia quaesitor pelleyorum, Demansia quaesitor garrodi, Demansia shinei starkeyi, Vipera hoserae, Vipera wellsi, Vipera wellingtoni, Vipera britoi, Vipera veloantoni, Amphibolurus jacky, Amphibolurus eipperi, Amphibolurus adelyn, Amphibolurus wellsei, Lophognathus wellingtoni, Melvillesaurea, Notactenophorus, Pseudoctenophorus, Chapmanagama, Turnbullagama, Paractenophorus, Leucomaculagama, Arenicolagama, Valenagama, Aurantiacoagama, Membrumvariegatagama, Rankina hoserae, Rankinia jameswhybrowi, Rankinia neildaviei, Rankinia fergussonae, Pailsagama, Diporiphora melvillae, Diporiphora smithae, Diporiphora shooi, Diporiphora harmoni, Diporiphora nolani, Diporiphora garrodi, Tympanocryptis markteesi, Tympanocryptis alexteesi, Tympanocryptis bottomi, Platyelapid, Salmonelaps par choiseulensis, Salmonelaps par ngelaensis, Salmonelaps par shortlandensis, Salmonelaps desburkei, Loveridgelaps sloppi, Loveridgelaps josephburkei, Loveridgelaps yeomansi, Loveridgelaps fiacummingae, Xenodermus oxyi, Xenodermus crottyi, Xenodermus sloppi, Fereachalinus, Parastoliczkia, Tropidechis jessejacksoni, Candoia hoserae, Candoia simonmcgoldricki, Candoia louisemcgoldrickae, Candoia jamiekonstandinoui, Candoia jamiekonstandinoui georgekonstandinoui, Candoia woolfi, Candoia kimmooreae, Candoia malcolmmclurei, Candoia boutrosi, Candoia niraikanukiwai, Candoia georgemacintyrei, Denisonia gedyei,* Carphodactylini, Nephruriini, Uvidicolina, Nephruriina, *Quazinephrurus, Paranephrurus, Nephrurus,*

Orrayini, Shireengeckiini, Oxygeckoina, *Oxygecko, Couperus, Teesgecko,* Shireengeckiina, *Saltuarius, Quazisaltuarius, Shireengecko, Quazishireengecko, Carphodactylus hoserae, Uvidicolus covacevichae, Saltuarius adelynae, Saltuarius jackyae, Nephrurus blacki, Nephrurus coreyrentoni, Nephrurus ianrentoni, Underwoodisaurus perthensis, Underwoodisaurus mensforthi, Underwoodisaurus mensforthi martinekae, Nephrurus levis bulliardi, Nephrurus sheai kimberleyae, Nephrurus asper saxacola, Corucia hoserae, Corucia woolfi, Corucia elfakhariorum, Quazitribolonotus, Feretribolonotus, Propetribolonotus, Quazitribolonotus frankanthonyi, Quazitribolonotus tomlonsdalei, Feretribolonotus greeri, Tribolonotus gracilis karkarensis,* Tribolonotiini, Fojiina, Tribolonotiina, *Adelynhosersaur spinipes adelynae, Adelynhosersaur spinipes jackyae, Adelynhosersaur spinipes wilkiei, Hypsilurus boydii ruivenkamporum, Gnypetoscincus smythi, Daraninagama robinsonii cliveevatti, Lazarusus, Boiga irregularis halmaheraensis, Boiga irregularis roddai, Boiga irregularis buruensis, Boiga irregularis sudestensis, Boiga irregularis solomonensis, Boiga irregularis newbritainensis, Apexvipera, Montivipera europa, Montivipera snakebustersorum, Montivipera yeomansi, Euanedwardssaurus, Hoplocephalus stephensi boutrosi, Hoplocephalus stephensi andrewgedyei, Emydocephalus teesi, Sayersus funki, Sayersus idyllwildi, Uropsophus gedyei, Uropsophus rentoni, Aplopeltura daranini, Aplopeltura gibbonsi, Aplopeltura lynnejohnstoneae, Aplopeltura omarelhelou, Aplopeltura shireenae, Pareas malayensis, Pareas sumatrensis, Litotescincus wellsi, Adelotus griffithsi, Adelotus valentici, Rotundishus, Platyplectrum shaneblacki, Rotundishus hayi, Paramixophyes, Mixophyes couperi, Mixophyes shireenae, Paramixophyes piersoni, Paramixophyes yeomansi,* Fiacumminggeckoini, Fiacumminggeckoina, Celertenuina, Hesperoedurina, Nebuliferina, *Strophuriina, Fiacumminggecko, Marlenegecko, Robwatsongecko, Fereoedura, Celertenues, Oedura bulliardi, Oedura bulliardi whartoni, Oedura bulliardi bulliardi, Oedura rentonorum, Fiacumminggecko fiacummingae, Fiacumminggecko richardwellsi, Fiacumminggecko rosswellingtoni, Fiacumminggecko charlespiersoni, Fiacumminggecko matteoae, Fiacumminggecko dorisioi, Fiacumminggecko julianfordi, Marlenegecko shireenhoserae, Celertenues bobbottomi, Celertenues evanwhittoni, Celertenues helengrasswillae, Amalosia alexanderdudleyi, Marlenegecko tryoni eungellaensis, Marlenegecko tryoni davidcharitoni, Nebulifera robusta merceicai, Adelyndactylus, Graciledactylus, Parvusdactylus, Oedurella alba, Oedurella garystephensoni, Oedurella jamielindi, Oedurella sonnemanni, Oedurella taeniata minima, Strophurus chriswilliamsi, Strophurus dannybrowni, Strophurus gedyei, Strophurus jackyae, Strophurus jenandersonae, Strophurus intermedius obscurum, Crottyoides, Lucasium allengreeri, Lucasium rosssadlieri, Honlamopus, Gracileopus, Crottyopus, Sloppopus, Wellingtonopus, Wellsopus, Brettbarnettus, Maryannmartinekea, Delma megleesae, Pseudodelma cummingae, Crottyopus jamesbondi, Wellingtonopus stevebennetti, Wellingtonopus grahamrichardsoni, Wellsopus shanekingi, Wellsopus brianbarnetti, Wellsopus kylienaughtonae, Wellsopus michaelguiheneufi, Wellsopus richardwarneri, Wellsopus robwatsoni, Pygopus brettbarnetti, Pygopus woolfi,* Aprasiaini, Aprasiaina, Pletholaxina, Ophidiocephalina, Pygopusini, Pygopusina, Paradelmaina, Lialisini, Sloppopini, Sloppopina, Crottyopina, *Boiga donagheyae, Boiga germainegreerae, Boiga mickpitmani, Ruivenkamporumus, Elfakhariorumserpens, Matsonserpens, Telescopus mannixi, Telescopus gocmeni, Swilearanea, Shireenaranea, Grayaraneaus, Wongaraneaus, Aspidoclonion sloppi, Megaerophis flaviceps promontoriumrursus, Megaerophis flaviceps masalbidus, Calloselasma oxyi, Morelia cliffrosswellingtoni, Fitzroychelys, Elseya shireenhoserae, Elseya albagula fitzroyi, Novamyuchelys, Wellsandwellingtonchelys, Magdelenachelys, Erythrocephalachelys, Crottyopus scottmarshalli, rottyopus daveausteni, Wellingtonopus matthingleyi, Fiacummingea, Woolfscincus, Piersonsaurus, Mannixsaurus, Silubosaurus hoserae, Silubosaurus hoserae hoserae, Silubosaurus hoserae maxinehoserae, Woolfscincus maryannmartinekae, Woolfscincus halcoggeri, Contundo rosswellingtoni, Silubosaurus zellingi fiacummingae, Silubosaurus zellingi scottgranti, Silubosaurus zellingi doriskuene, Silubosaurus stokesii lynetteholdworthae, Mannixsaurus formosa matthingleyi, Silvascincus richardi adrianpapalucai, Flamoscincus kintorei crossi, Flamoscincus kintorei crossmani, Gymnobelideus leadbeateri martinekae, Oopholis oxyi, Aechmophrys adelynhoserae, Aechmophrys jackyhoserae, Elachistodon westermanni dannybrowni, Feresuta, Feresuta hamersleyensis, Feresuta monachus centralis, Feresuta monachus interiorensis, Hulimkai punctata divergens, Hulimkai fasciata ruber, Worrellisaurus kimaniadilbodeni, Worrellisaurus microocellata, Worrellisaurus tyeseeipperae, Worrellisaurus scotteipperi, Worrellisaurus dannybrowni, Worrellisaurus jenandersonae, Worrellisaurus bigmoreum, Odatria davidhancocki, Odatria glebopalma jimgreenwoodi, Halmaherasaurus, Purpuracolotes, Maculocolotes, Wedgedigitcolotes, Saxacolinecolotes, Propemaculosacolotes, Crocodilivoltuscolotes, Edaxcolotes, Macrocephalacolotes, Extensusdigituscolotes, Brevicaudacolotes, Parvomentumparmacolotes, Papuacolotes, Quattuorunguiscolotes, Colotesmaculosadorsum, Thaigehyra, Gehyra hangayi, Phryia paulhorneri, Dactyloperus bradmaryani, Dactyloperus bradmaryani bulliardi, Extensusdigituscolotes sadlieri, Extensusdigituscolotes glennsheai, Crocodilivoltuscolotes shireenhoserae, Crocodilivoltuscolotes marleneswileae, Dactyloperus spheniscus graemecampbelli, Dactyloperus federicorossignolii, Quattuorunguiscolotes grismeri, Liopholis dannygoodwini, Sparsuscolotes, Sinogekko, Lautusdigituscolotes, Aurumgekko, Glanduliscrusgekko, Magnaocellus, Extentusventersquamus, Cavernagekko, Foderetdorsumgekko, Alexteescolotes, Cliveevattcalotes, Alexteescolotes teesi, Cliveevattcolotes steveteesi, Ptychozoon cliveevatti, Ptychozoon sumatraensis, Ptychozoon malayaensis, Ptychozoon johorensis, Ptychozoon engannoensis, Ptychozoon sulawesiensis, Ptychozoon borneoensi, Ptychozoon wallaceaensis, Borealiscolotes, Shireenhosergecko, Jackyhosergecko, Solomoncolotes, Bobbottomcolotes, Martinekcolotes, Adelynhosergecko, Allengreercolotes, Borneocolotes, Rosssadliercolotes, Charlespiersoncolotes, Haroldcoggercolotes, Georgemarioliscolotes, Robwatsoncolotes, Shireenhosergecko shireenhoserae, Shireenhosergecko petewhybrowi, Shireenhosergecko robjealousi, Shireenhosergecko dalegibbonsi, Shireenhosergecko jarradbinghami, Jackyhosergecko jackyhoserae, Bobbottomcolotes bobbottomi, Bobbottomcolotes potens, Bobbottomcolotes crusmaculosus, Allengreercolotes allengreeri, Allengreercolotes pauldarwini, Allengreercolotes paulwoolfi, Adelynhosergecko adelynhoserae, Adelynhosergecko sloppi, Adelynhosergecko huonensis, Adelynhosergecko madangensis, Adelynhosergecko judyfergusonae, Adelynhosergecko haydnmcphiei, Adelynhosergecko matteoae, Adelynhosergecko stevebennetti, Adelynhosergecko lucybennettae, Adelynhosergecko lachlanmcpheei, Scelotretus haroldcoggeri, Scelotretus daranini, Scelotretus jenandersonae, Paraheleioporus, Philocryphus hoserae, Ferehemiphyllodactylus, Maculacruscalotes, Cassandracampbellea, Malayacolotes, Titiwangsacolotes, Malayacolotes cassandracampbellae, Chelosania neilsonnemanni, Tympanocryptis snakebustersorum, Tympanocryptis optus, Tympanocryptis vodafone, Tympanocryptis lachlanheffermani, Tympanocryptis tetraporophora iarentoni, Tympanocryptis simonknolli, Tympanocryptis simonknolli marcusbrummeri, Tympanocryptis deniselivingstonae, Tympanocryptis karimdaouesi, Tympanocryptis karimdaouesi courtneyleitchae, Tympanocryptis williamconnellyi, Tympanocryptis tonylovelinayi, Tympanocryptis reconnectorum, Tympanocryptis reconnectorum clintonlogani, Tympanocryptis samsungorum, Williamconnellysaurus, Notanemoia, Cannotbeemoia, Silvaemoia, Griseolaterus, Aintemoia, Paraemoia, Caeruleocaudascincus, Ventripallidusscincus, Aquilonariemoia, Shireenhoserscincus, Emoia kimaniadilbodeni, Emoia timdalei, Emoia karkarensis, Emoia tonylovelinayi, Emoia anggigidaensis, Emoia stefanbroghammeri, Emoia euanedwardsi, Emoia martinmulvanyi, Emoia paulmulvanyi, Emoia paulwoolfi, Emoia jamesbondi, Emoia richardwarneri, Emoia morriedorisioi, Emoia minusguttata, Emoia dorsalinea, Emoia bougainvilliensis, Emoia boreotis, Emoia aquacauda, Emoia yusufmohamudi, Emoia davidaltmani, Emoia stephengoldsteini, Emoia rodneysommerichi, Emoia roberteksteini, Notanemoia georgemariolisi, Notanemoia karlagambellae, Notanemoia cathysonnemannae, Notanemoia neilsonnemanni, Aintemoia dannygoodwini, Aintemoia latishadarwinae, Caeruleocaudascincus stevebennetti, Caeruleocaudascincus lucybennettae, Caeruleocaudascincus clivebennetti, Caeruleocaudascincus craigbennetti, Caeruleocaudascincus brettbarnetti, Caeruleocaudascincus williambennetti, Caeruleocaudascincus kamahlbenneti, Caeruleocaudascincus drubennetti, Caeruleocaudascincus jaibennetti, Caeruleocaudascincus danielbennetti, Ventripallidusscincus graysonoconnori, Aintemoia michaelguiheneufi, Aintemoia rosssadlieri, Silvaemoia robvalentici, Shireenhoserscincus shireenhoserae, Shireenhoserscincus daranini, Dendrolagus hoserae, Dendrolagus ursinus arfakensis, Vermicella multifasciata kimberleyensis, Vermicella annulata paulmulvanyi, Vermicella annulata isaensis, Caimanops amphiboluroides aurantiaco, Caimanops amphiboluroides leucolateralis, Greersaurus, Paragreersaurus, Barnettsaurus, Kerryleewennigea, Coggersaurus, Tropidophorus joeymontebelloi, Aspris peterkraussi, Aspris russellgranti, Asiascincella, Ovipacincella, Sinoscincella, Ferescincella, Quaziscincella, Divergesaurus, Ferescincella insignipicturaconlus, Ferescincella ventrealbis, Ferescincella flavolateralis, Ferescincella yonagunijimaensis, Asymblepharus aurisovalibus, Asymblepharus lateralibusdorsoclavo, Crottysaurus, Retroalbascincus, Lateratenebriscincus, Pointednasus, Viridihaema, Macrotympanoscincus, Variusscincus, Crudushaema, Fojia aurantiacocauda, Lobulia oliveetfatua, Crottysaurus crottyi, Lateratenebriscincus sentaniensis, Lateratenebriscincus sepikensis, Lateratenebriscincus albaaudere, Lateratenebriscincus gulagorum, Lateratenebriscus tokpisinensis, Lateratenebriscus freshsweetpotato, Lateratenebriscus acrilineata, Lateratenebriscus maculaoccipitalis, Lateratenebriscincus albavarietata, Lateratenebriscincus laterafusca, Lateratenebriscincus etfatubrunnea, Lateratenebriscus leucolabialis, Pointednasus widerecta, Pointednasus clavoflavoviridis, Pointednasus currearbor, Pointednasus flavorecta, Pointednasus flavopalpebrae, Pointednasus ventriiridescens, Pointednasus makiraensis, Pointednasus extentadigitus, Pointednasus labiamarmorata, Variusscincus litoresaurus, Crudushaema allengreeri, Crudushaema haroldcoggeri, Rankinia hoserae martinekae, Burramys parvus hosersbogensis, Burramys parvus timdalei, Burramys parvus scottyjamesi, Cercartetus hoserae, Eudromicia richardwellsi, Eudromicia adelynhoserae, Eudromicia rosswellingtoni, Eudromicia jackyhoserae, Eudromicia doriskuenae, Sloppossum, Pseudochirops fiacummingae, Pseudochirops jamesbondi, Pseudochirops waddamaddawidyu, Pseudochirops archeri chrismaxwelli, Rossignolius, Potorous waddahyamin, Petaurus australis adelynhoserae, Quasipetrogale, Ferepetrogale, Petrogale hoserae, Petrogale martinekae, Petrogale brachyotis pentecostensis, Petrogale brachyotis ordensis, Ctenophorus adelynhoserae, Ctenophorus jackyhoserae, Ctenophorus katrinahoserae, Ctenophorus lenhoseri, Ctenophorus maxinehoserae, Ctenophorus ronhoseri, Ctenophorus sharonhoserae, Ctenophorus shireenhoserae, Limnodynastes alexantenori, Limnodynastes cameronganti, Limnodynastes shanescarffi, Platyplectron gerrymarantellii, Platyplectron timjamesi, Ranaster snakemansbogensis, Ranaster henrywajswelneri, Ranaster scottyjamesi, Ranaster lignarius divergens, Oxyslop, Feremixophyes, Quasimixophyes, Mixophyes hoserae, Hoserranae, Scottyjamesus, Hoserranae acutirostris shaunwhitei, Scottyjamesus rheophilus scottyjamesi,* Oxyslopidae, Oxyslopinae, Oxyslopini, Hoserranidae, Hoserraninae, Hoserranini, Scottyjamesini, *Macquaria hoserae, Macquaria honlami, Eulamprus paulwoolfi, Intellagama wellsandwellingtonorum, Ctenophorus fordi scottgranti, Ctenophorus fordi danielmani, Ctenophorus fordi scottyjamesi, Ctenophorus hawkeswoodi mary-*

annmartinekae, Brachyurophis alexantenori, Brachyurophis paultamisi, Brachyurophis paulwoolfi, Brachyurophis lesshearimi, Brachyurophis richardshearimi, Narophis richardwellsei, Narophis cliffrosswellingtoni, Simoselaps fukdat, Adelynhoserhyleini, *Yikesanura, Adelynhoserhylea, Adelynhoserhylea adelynhoserae, Adelynhoserhylea yikes, Jackyhoserhylea, Jackyhoserhylea ernieswilei, Jackyhoserhylea jackyhoserae,* Leucodigiranina, *Leucodigirana, Euscelis booroolongensis dorsaruber, Euscelis booroolongensis occultatum,* Coggerdoniani, Cycloraninini, *Mitrolysis alboguttata dumptrashensis, Paramitrolysis, Mitrolysis verrucosa inornata, Invisibiliaauris, Mitrolysis flavoranae, Mitrolysis leucodorsalinea, Neophractops rosea, Crottyanura, Crottyanura crottyi,* Ranoideina, *Sandgroperanura, Chirodryas sloppi,* Gedyeranina, *Gedyerana, Gedyerana gedyei, Amnisrana, Mosleyia cottoni, Mosleyia michaelsmythi, Mosleyia pilloti,* Daraninanurini, *Daraninanura,* Fiacumminganurini, *Fiacumminganura, Fiacumminganura fiacummingae, Fiacumminganura timdalei,* Dryopsophina, *Leucolatera, Ausverdarana, Dryopsophus citropa gippslandensis, Dryopsophus jarrodthomsoni,* Kumanjayiwalkerini, *Kumanjayiwalkerus, Kumanjayiwalkerus kumanjayi,* Audaxurina, *Audaxura, Brevicrusyla, Balatusrana, Colleeneremia bogfrog, Colleeneremia chunda, Colleeneremia dunnyseat, Colleeneremia watdat, Colleeneremia wifi, Colleeneremia dentata toowoombaensis,* Rawlinsonina, *Rawlinsonia ventrileuco,* Maxinehoserranini, *Maxinehoserranae, Vegrandihyla, Maxinehoserranae brettbarnetti, Maxinehoserranae maxinehoserae, Maxinehoserranae piersoni, Angularanta, Alliuma, Longuscrusanura, Naveosrana, Raucus, Scelerisqueanura, Angularanta chydaeus, Angularanta communia, Angularanta extentacrus, Angularanta mukherjii, Angularanta quaeinfernas, Angularanta vulgarans, Angularanta lutea leucopunctata, Angularanta oxyeei, Angularanta louisiadensis brunetus, Bellarana, Fluvirana, Hopviridi, Incertanura, Incertanura cuspis, Incertanura fakfakensis, Inlustanura, Inlustanura inluster, Moechaeanura, Aspercutis, Telaater, Moechaeanura tritong, Moechaeanura albatermacula, Moechaeanura spica, Ornatanura, Ornatanura leucopicturas, Ornatanura parscinereo, Ornatanura parsviridus, Nasuscuspis, Rotundaura, Variabilanura, Sudesanura, Variabilanura tomcottoni,* Drymontantina, *Drymomantis ausviridis, Drymomantis celantur, Drymomantis northstradbrokensis,* Nyctimystini, *Magnummanibus, Asperohyla, Nyctimystes mon-*

doensis, Nyctimystes charlottae, Nyctimystes doggettae, Nyctimystes aspera, Nyctimystes georgefloydi, Albogibba, Ratiobrunneis, Albogibba ingens, Occultatahyla, Webpede, Nigreosoculus, Badiohylina, *Badiohyla, Magnumoculus,* Pelodryanini, Shireenhoserhylina, *Shireenhoserhylea, Emeraldhyla, Shireenhoserhylea megaviridis, Summaviridis,* Pustulataranini, *Pustulatarana, Pustulatarana longirostris tozerensis, Microlitoria, Llewellynura fukker, Llewellynura yehbwudda, Mahonabatrachus chriswilliamsi, Mahonabatrachus marionanstisae, Mahonabatrachus pailsae, Mahonabatrachus roypailsi,* Salmocularanina, *Salmocularana, Salmocularana saxacola, Paralitoria, Ferelitoria, Vultusamolitoria, Quasilitoria inermis davidtribei, Quasilitoria inermis dunphyi, Quasilitoria mickpughi, Quasilitoria mippughae, Quasilitoria tornieri serventyi,* Sagunurini, Wowranaini, *Wowrana, Parawowrana,* Sandyranina, *Euprepiosaurus adelynhoserae, Euprepiosaurus allengreeri, Euprepiosaurus dorisioi, Euprepiosaurus elfakhariorum, Euprepiosaurus jackyhoserae, Euprepiosaurus lenhoseri, Euprepiosaurus matteoae, Euprepiosaurus oxyi, Euprepiosaurus paulwoolfi, Euprepiosaurus powi, Euprepiosaurus scottgranti, Euprepiosaurus sloppi, Shireenhosersaurea clara, Shireenhosersaurea satis, Shireenhosersaurea shireenhoserae, Parvavaranus apicemalba, Parvavaranus ignis, Parvavaranus pyrrhus, Vermicella sloppi, Zilonear, Dannyleeus danielmani, Dannyleeus rayhammondi, Dannyleeus tongzhoujiae, Katrinahoserserpenea bobbottomi, Katrinahoserserpenea danielmannixi, Katrinahoserserpenea daranini, Katrinahoserserpenea evanwhittoni, Katrinahoserserpenea mcconnachiei, Katrinahoserserpenea rodneykingi, Assa brianchampioni, Assa guatammukherjii, Assa jamesbondi, Assa roberteksteini, Crotalus wellsi,* Crotalus *wellingtoni, Uropsophus oxyi, Uropsophus armstrongi strimplei, Uropsophus woolfi, Uropsophus euanedwardsi, Uropsophus elfakhariorum, Uropsophus valentici, Uropsophus swileorum, Uropsophus aquilus hammondi, Cottonus tomcottoni, Matteoea mitchelli matteoae, Matteoea pyrrhus dorisioi, Matteoea stephensi sommerichi, Piersonus brunneus bartletti, Caudisona evatti, Caudisona basiliscus teesi, Caudisona molossus* smythi, *Ophioscincus paulwoolfi, Jackyhosersaur superba jackyhoserae, Shireenhoserhylea shireenhoserae, Chondropython viridis jackyhoserae, Acanthophis oxyi, Acanthophis crotalusei karkarensis, Macrocalamus wellsei, Macrocalamus wellingtoni, Collorhabdium daranini, Oreocalamus turneri, Raclitia oxyi,*

Paraisopachys, Isopachys roulei rosswellingtoni, Tropidonotus truncatus morotaiensis, Eekmys, Farkmys, Ohmys, Oimys, Ouchmys, Pseudomys albapes, Pseudomys griseorursus, Pseudomys pesrosea, Pseudomys pellicauda, Pseudomys johnsoni luxauris, Pseudomys johnsoni occultatum, Macropogonomys, Macropogonomys maxhoseri, Macropogonomys maxhoseri blacki, Macropogonomys maxhoseri gedyei, Macropogonomys mickpughi, Macropogonomys mippughae, Macropogonomys aplini, Pogonomys sharonhoserae, Bogophryne, Bogophryne uterbog, Bogophryne duboisi, Bogophryne naomiosakaae, Sloppophryne, Crottyphryne, Oxyphryne, Crottyphryne crotalusei, Crottyphryne douglasi oxyi, Kankanophryne maxinehoserae, Kankanophryne katrinahoserae, Kankanophryne marcdorsei, Bufonella hoserae, Bufonella hoserae sadlieri, Bufonella woolfi, Bufonella euanedwardsi, Pseudophryne scottgranti, Pseudophryne jasminegranti, Pseudophryne martinekae, Pseudophryne wellsi, Pseudophryne wellingtoni, Pseudophryne wellingtoni kupulurensis, Pseudophryne dendyi mensforthi, Pseudophryne semimarmorata burrelli, Oxyodella, Crinia oxeyi, Crinia crottyi, Crinia sloppi, Lowingdella, Crinia lowingae, *Crinia stevbennetti, Crinia maateni, Crinia merceicai, Wellingtondella, Geocrinia brettbarnetti, Geocrinia brianbarnetti, Geocrinia laevis grampiansensis, Geocrinia victoriana otwaysensis, Geocrinia victoriana logani, Paracrinia lenhoseri, Paracrinia funki, Metacrinia bettyswileae, Metacrinia wilhelminahughesae, Quasiuperoleia, Hosmeria shuddafakup, Hosmeria shireensbogensis, Uperoleia jadeharrisae, Uperoleia keilleri, Uperoleia lowryi, Uperoleia shanescarffi, Uperoleia micra divergans, Uperoleia margweeksae, Uperoleia margweeksae maximus, Uperoleia grantturneri, Uperoleia minima dispar, Uperoleia gedyei, Uperoleia rossignolii, Mixophyes hoserae jackyae,* Myobatrachini, Oxyphryneina, Spicospinaina, Uperoleiaina, Paracriniaina, Criniaini, Wellingtondellaini, *Chelydra haydnmcphiei, Martinekchelys, Parasternotherus, Crottychelys, Crottychelys perixantha ipsumtenebris, Oxychelys, Oxychelys oxyi, Kinosternon baurii grantturneri, Sloppchelys, Clemmys guttata maximus, Clemmys guttata praetortus, Actinemys maxinehoserae, Heosemys turneri, Hieremys grandis malayensis, Hieremys annandalii mekongensis, Vijayachelys silvatica whittoni, Chitra indica indusensis, Parageoemyda, Geoemyda daranini, Freudchelys, Freudchelys freudi, Chelonoidis hoserae, Chelonoidis woolfi, Chelonoidis fiacummingae, Parachelonoidis, Kinixys homeana varians, Kinixys erosa divergentens, Cyclemys mcdermottorum, Graptemys caglei flavoculus, Graptemys pseudogeographica brunneisoculus, Graptemys geographica aurantiacooculus, Emys orbicularis repens, Cuora adelynhoserae, Cuora jackyhoserae, Cuora oxyslopp, Cuora boxboyi, Cuora elfakhariorum, Cuora richardwellsi, Cuora rosswellingtoni, Chersina swileorum, Chersobius mandela, Funkichelys, Funkichelys funki, Homopus trevorhawkeswoodi, Homopus trevorhawkeswoodi knysaensis, Homopus trevorhawkeswoodi bloemfontainensis, Cyclanorbis senegalensis nileensis, Cyclanorbis senegalensis occultatum, Cycloderma tismorum, Heptathyra marcdorsei, Piersonchelys, Parapelodiscus, Amyda ashphillipsi, Amyda ornata magnapapulae, Keillerchelys, Pelomedusa darrenkeilleri, Pelomedusa alexstaszewskii, Pelomedusa dannygoodwini, Pelomedusa shannonmcgrathi, Pelusios lynnrawi, Pelusios rhodesianus divergentans, Orlitia borneensis perakensis, Emydura wellingtoni, Emydura wellsi, Emydura hawkeswoodi, Wollumbinia georgefloydi, Wollumbinia darnellafrazierae, Hydromedusa meyeyouchelys, Wittchelys, Wittchelys tectifera wittorum, Lovelinaychelys,* Tetenditunguini, *Parapuellula, Geckoella wijenayakai, Geckoella rowvillewilsonorum, Tetenditunguis, Tetenditunguis hoserae, Tetenditunguis gedyei, Tibetgekko, Aeschtgekko, Ahyonggekko, Elegansgekko, Balleriogekko, Bertlinggekko, Purpurabrunusgekko, Bouchardgekko, Bourgoingekko, Dmitrievgekko, Dmitrievgekko oxyi, Evenhuisgekko, Exilgekko, Grygiergekko, Claragekko, Kottelatgekko, Tenuisalbavincula, Cuprumcinctim, Kullandergekko, Terrenusgekko, Arenosumgekko, Kullandergekko rosswellingtoni, Kullandergekko richardwellsi, Kullandergekko trevorhawkeswoodi, Kullandergekko wennigae, Papegekko, Triavincula, Lyalgekko, Macrolyalgekko, Pylegekko, Rheindtgekko, Rosenberggekko, Melanogekko, Rosenberggekko caudatenuis, Rosenberggekko merceicai, Rosenberggekko alexanderdudleyi, Rosenberggekko fritzmaarteni, Welterschultesgekko, Welterschultesgekko keilleri, Welterschultesgekko scottgranti, Zhiqiangzhanggekko, Maculagekko, Morotaigekko, Infigo, Infigo jackyhoserae, Caudacingitur, Russetocolore, Nigricansalvum, Facileoccultatur, Purpuraoculus, Fasciacorpus, Linguarosea, Albatubercula, Graysongekko, Vinculatigris, Obscuramacula, Spinagekko, Maculatumetglobum, Tuberculatasinus, Fasciaincompletum, Arunachalgekko,* and *Brunneisoculura.*

By mid-October 2021, Hoser had described nearly a thousand new species and subspecies, along with another thousand genera, tribes, and so on. One of every ten scientific names that scientists had given reptiles over the past 270 years was Hoser's. The next most prolific namer of reptiles, Boulenger, had named a mere 659 species and subspecies. Cope, the next most prolific after him, named 385. Günter, the third, named 364, and Gray, the fourth, 335. Among taxonomists still living, the most prolific namer of reptiles after Hoser is Bauer, who has named a paltry 164 species and subspecies.

But even as the International Commission on Zoological Nomenclature rejected efforts by Hoser's opponents to censure him using the nomenclatural code, Hinrich Kaiser's boycott had taken hold. For years, Hoser's voluminous taxonomic efforts had drawn argument and denunciation and hand-wringing from across the globe. Now there was only silence.

CHAPTER 11

BABEL

In various moulds and frames all things were cast,
But none forever can endure nor last.
Whatever took a form, must change or mend;
Whatever once began, must have an end.

—CONSTANTINE RAFINESQUE, *THE WORLD, OR INSTABILITY*

There was one more stop. We had to pick up Hoser's new prescription sunglasses. "Ah, there's a newsstand," he said, pointing to a stack of *The Vision China Times*, a Chinese-language paper, as we weaved through a crowded mall. Then he returned to explaining his ongoing trademark-related lawsuits against various rival snake men and Bunnings, the Australian hardware-store chain, which he said was connected in surprising ways to the Wüster Gang. While we waited in line at the glasses shop, he took off one of his boots and banged it out, depositing a generous quantity of dirt and twigs on the shop floor. He recounted how he had discovered a certain Indian turtle species that was distinguishable from other, similar turtles not only by its distribution and genetics but also by its eyes, which were invariably blue. It was one of the turtles that the Wüster Gang's allies in the Turtle Taxonomy Working Group claimed he had taxonomically stolen from them, he said, which was clearly untrue, evidenced in part by the middling quality of the blue-eyed turtle itself. "Let's assume I know nothing about turtles," he said. "If I was going to start picking turtles to name, you'd think I'd start naming better ones."

We left the glasses shop and went deeper into the mall, opposite the way we'd entered. We came to a grocery store. Hoser searched for a cart. There was a mysterious shortage. Finally he found one and started to push it away from the grocery store before discovering it had a bad wheel. He abandoned it and spotted another cart thirty feet away. A woman with two small children had sighted it at the same time. A standoff ensued. "Grab that one," Hoser suggested, pointing at the cart with a bad wheel. She gave a nervous laugh. Another second passed before Hoser relented. "Now, see, that's the difference between me and Wüster," he told me, as we sped off in another direction through the mall. "If that had been Wüster, he would've fought and taken that trolley. Me, if someone else says, 'I want to name that animal,' I say, 'Go ahead.' I would never jump in front of someone who's about to name an animal. Wüster would've fought for that trolley."

At a second market, he found a working cart. He pushed it back the way we had come. We passed the first grocery store and the glasses shop and stopped next to the stack of *The Vision China Times*. The stand was labeled, in English, "One per customer." He started putting the newspapers in the cart. "Because they're free, they can't bust you for stealing the newspaper," he said. A man of about Hoser's age was leaning on a railing nearby, and as Hoser loaded armful after armful of newspapers into the cart, a look of bafflement and then of indignation spread across the man's face. I avoided his gaze. When Hoser had gathered up every last newspaper, he pushed the shopping cart back toward the parking garage.

This is what it was like to hang out with Raymond Hoser. He was Woland in Bulgakov's Moscow, Mephistopheles with Gauthe's Faust, the snake in Milton's Garden—a force of chaos, mischief, rebellion. He did not seem to perceive the invisible fences that corral most members of society. A man in a tiny black hat, he galloped freely across the experiential range. To see someone so unencumbered was bracing at times, alarming at others. It was always a spectacle. People gawked and stared. Hoser, to the degree he noticed this attention, seemed to enjoy it. He was never at a loss for words, relying on an arsenal of jokes and conjunctional phrases like "so anyway," and "so what happened was," that allowed him to clear conversational bumps and hurdles and return fluidly to previous topics. Along for the ride, I often felt less like an observer and more like an accomplice. It was kind of fun.

Not everyone appreciated Hoser's jokes, or the way he swung their children around by their tiny extremities. After rescuing a man and his family from what turned out to be a harmless blue-tongue lizard, Hoser (correctly) identified the man's child as a lockdown baby and offered his Snakebusters business card. The man looked down at the card but did not reach to take it. "I hope never to see you again," he told Hoser.

But many others seemed to delight in his antics, to find them admirable, even in the retelling. My doctor, for instance, after listening with a look of bored sympathy while I described the symptoms of some now-forgotten ailment, perked up when I started telling him about Raymond Hoser the Australian Snakeman's method of discovering species on the branch tips of other scientists' phylogenetic trees, a method that, I explained, the other scientists tended to see as taxonomic theft.

"Good for him!" my doctor said.

It seems Hoser hums a universal tune. "Perhaps all but the extremely privileged in wealth and power in any civilized society yearn subconsciously for freedom from the constraints of civilization," wrote Paul Angiolillo in *A Criminal as Hero*. "Routine, duty, responsibility, and dependency have been (and continue to be) the inevitable fruits of organized society. Both peasant and rich man, young and old, can look upon the devil-may-care type as an enviable example of one who had the courage to break out of the mold."

But where some see the romance of a life in the greenwood, others see missed potential. Nearly all of the taxonomists and herpetologists I spoke with about Hoser's taxonomic efforts noted his intelligence and energy and the depth of his experience with reptiles. Many suggested he could have made useful contributions to herpetology and even to the collective task of taxonomy, were he not so consumed by a spirit of rebellion. Indeed, in his decades-long effort to uncover a conspiracy observable only by him, he seems to have made it a reality. After the International Commission on Zoological Nomenclature declined to censure Hoser in 2021, Wolfgang Wüster, Hinrich Kaiser, Scott Thomson, and Mark O'Shea published a paper announcing their intention to continue the boycott of Hoser's names despite the lack of support from the commission. "Zoologists have not only a right but indeed a duty to uphold the principles of science against malicious, unscientific taxonomic work," they wrote, "preferably within the letter of the *Code*, but, with deep regret and only as a last resort, outside it if necessary." The letter was signed by some 458 scientists in fifty-three countries. In their telling, the world's herpetologists and taxonomists were in fact acting in the orderly, law-abiding spirit of Linnaeus, of Hugh Strickland, setting aside their individual differences and agreeing to a common cause—in this case, suppressing the rebel in their midst. When we first spoke in early 2023, Kaiser, the boycott's architect, told me the Hoser problem had been solved. "I know of nobody who uses those names," he said. "Not one."

Hoser's interest in taxonomy and nomenclature, his attention to small legalistic details, his indifference to social norms, and his reflexive defiance seem like such a rare combination of traits that they would be unlikely to co-occur in anyone else. Yet it appears that, to the contrary, many taxonomists worry that there are numerous other rebellious souls who, given the opportunity, might launch their own taxonomic sprees—that Hoser represents merely a hint of what could be to come. "Taxonomic vandalism—the malicious dismantling of scientific classifications—has become the opiate of the unskilled and unsophisticated," wrote one researcher in a recent paper. In another paper, a group of scientists suggested that the failure of the International Commission on Zoological Nomenclature to censure Hoser would inspire imitators. "The result of these hesitations and uncertainties will doubtless be that such practices will spread," they wrote. Yet another scientist, in an email declining my invitation to talk about Hoser, wrote: "Anyone in the world could do what he has done. . . . We're lucky there aren't more copycats."

Wells and Wellington, in the 1984 paper in which they coined the phrase "taxonomic vandalism," used it as a kind of preventive measure, meant to preempt accusations against them. But naming is an act of creation, and what happened instead is that they summoned taxonomic vandalism into existence—if not as an actual phenomenon, then as an idea, a concept that other taxonomists could discuss, study, recognize, fear. And in taxonomists' fear of taxonomic vandalism, they have begun to see it all over.

In a series of recent papers, Geiger accused Chen, George, Averyanov, and others of taxonomically vandalizing the *Oberonia* orchids. Loizides et al. denounced taxonomic vandals working on the true morels. Marin-Felix and Miller lamented the taxonomic vandalism wreaked by Huang et al. upon the sac fungi. Moore, Jameson, and Paucar-Cabrera accused Soula of taxonomically vandalizing the scarab beetles. Chung et al. wrote that the work of Ying on *Lysmachia* was a "self-published and poorly edited work of apparent taxonomic vandalism." Kok described a paper by Rivas et al. on the anacondas as "a digest of bad taxonomic practice," and pointedly referred to several works about taxonomic vandalism. Teimori et al. wrote that they considered the work of Freyhof and Yoğurtçuoğlu on the killifish "remorseless, unethical, and an example of taxonomic vandalism, and our view is shared by all of our colleagues with which we have discussed this issue." Members of the International Committee on Systematics of Prokaryotes declared that the revisions of Gupta et al. to the Mycoplasmatales bacteria were "unnecessary over-reach verging on taxonomic vandalism." Páll-Gergely, Hunyadi, and Auffenberg accused Thach and colleagues of committing taxonomic vandalism in their description of some 235 new land snails; in response, Thach et al. maintained that the land snails they described were real and advised that constructive criticism should avoid "personal attacks, insults, and the use of offensive words like 'vandalism.'" The editors of *Koleopterologische Rundschau* decried the taxonomic vandalism wrought by Makhan in the beetles, ants, and bugs, and published a collection of excerpts from emails that Makhan had written to his professional colleagues. "Slime ball, you will be slimed," he wrote in one. "I was thinking you are a wise man, but you are so low," he wrote to another. "I shall publish many more papers, you will like it or not."

The boycott of Hoser's names seems to offer a route to simply ignoring the work of such colleagues rather than sorting through it, as required by the principle of priority, but Hinrich Kaiser and Wolfgang Wüster warned that other taxonomists should be cautious in beginning boycotts of their own. Every time someone ignores one of Hoser's names, "they're going against the International Code of Zoological Nomenclature," Wüster said. "At the end of the day, it is a code that everybody's supposed to follow. Going against it is a big deal." The hundreds of herpetologists who had signed onto the boycott against Hoser lent it legitimacy, he said. "You clearly wouldn't want just one or two people deciding they're going to override the Code and overwrite a name that they don't like."

That has happened on at least two recent occasions. One example came in 2019, when a group of five herpetologists led by Troncoso-Palacios cited Kaiser's boycott and announced that they would similarly ignore the contents of Demangel's self-published book *Reptiles en Chile*, in which Demangel included a number of taxonomic descriptions. In another paper, a team of herpetologists led by Esquerré described a genus of python that they called *Narawan*. Wells and Wellington had described the same python genus in their 1985 "Classification of the Amphibia and Reptilia of Australia," but the authors of this new work cited the boycott against Hoser and wrote that since "it is well understood that [Wells and Wellington's] self-published work does not adhere to good practices in taxonomy," they were justified in ignoring it. Kaiser and two colleagues wrote a response, arguing that the overwriting was not called for, and that the Wells and Wellington name should be used.

Some taxonomists predict this list will inevitably grow. The apparent shortcut offered by the boycott against Hoser set a dangerous precedent, opening a "Pandora's box," as one group of scientists put it, writing that "these practices have started to spread to other taxonomic domains, where various authors now decide to ignore deliberately some publications and even to apply the unclear expression 'taxonomic vandalism' to works that do not deserve this qualification in the least." As another scientist put it in an exchange with Wüster on a taxonomic message board, "It is precisely my worry that the lack of a clear dividing line between the extreme cases and the rest will make it possible to label genuine people as 'just another Hoser.'"

Indeed, there is no clear standard for what is and isn't taxonomic vandalism, and the question remains as to whether the phenomenon is truly new or is simply a rhetorical escalation. Taxonomists have been complaining about one another's work for centuries, after all, accusing one another of mania, malfeasance, and the mihi-itch, and there is no way of proving the ill intent implicit in the word *vandalism*; nobody, as far as I am aware, has ever admitted to being a taxonomic vandal. But it does not seem like a coincidence that the phrase should come to popularity in a societal moment characterized by the ascendance of anti-institutionalism, contrarianism, and suspicion of well-supported scientific conclusions of all kinds, by an apparent disdain for any kind of collective task. In such an atmosphere, little wonder that taxonomists should be sensitive to any sign of miscommunication, alert to anyone who might seem to betray their shared endeavor—that they should increasingly perceive one another as not just incompetent but malevolent.

In all of the articles and discussions around taxonomic vandalism, what motivates the alleged vandals goes mostly undiscussed. Perhaps it seems obvious. "A feeling of loneliness, alienation, mediocrity, and failure may trigger an envy directed against those perceived to be more successful or prestigious," writes Albert Borowitz in *Terrorism for Self-Glorification: The Herostratos Syndrome.* "Envy is exacerbated by an ambitious, competitive spirit and the conviction that avenues to success are unfairly blocked." He quotes Alberto Franceschini of the Italian Red Brigades: "When you do certain things, and these things turn into big paragraphs in the papers; and when you see that because of the things you explode, fights and chaos happens between the politicians; all this summed up gave me a sensation of great power." Sabato Rodia, builder of the Watts Towers in South Los Angeles, put it even more succinctly: "You have to be either good good or bad bad to be remembered."

The authors of Genesis offer the plainest, most matter-of-fact account of the Tower of Babel. When God saw what the people were building, He said, "Behold, the people is one, and they have all one language: and this they begin to do: and now nothing will be restrained from them, which they have imagined to do." So He scattered them and tore down their tower. As I thought about the story of Raymond Hoser's alleged acts of taxonomic vandalism and about the rebellion Kaiser and Wüster led against the International Code of Zoological Nomenclature, usually treated by taxonomists as nearly scriptural, I found myself imagining all the details missing from the story of Babel—the motivations and mechanisms, the betrayals and jealousies and egos. It began, I imagine, when the snake slipped in among the builders and began to whisper, causing them to fall into argument over their collective vision, sowing confusion over what everything should be called.

~

Surely part of the reason that taxonomists have become quick to deem one another taxonomic vandals, even to try to banish one another from the field, is the growing sense that they are running out of time. The world seems to contract before them, tipping deeper into an extinction crisis. Estimates of the speed and scale of this crisis vary widely, but many scientists think that the combined effects of habitat loss, globalization, and climate change could put tens or even hundreds of thousands of species at risk in coming decades. Every species is a unique evolutionary path, an expression of one way to survive, grow, and reproduce. With every extinction, the future course of life on Earth is forever altered. Less absolute but perhaps more immediately apparent is the reduction in the number of individual creatures. In 2024, the World Wildlife Fund announced that over the past fifty years, the populations of nearly 5,500 species under scientific monitoring fell by an average of more than 70 percent. Even the commonest of things seem threatened with rarity.

As life on Earth appears enfeebled, dimmer than before, its full design remains elusive. The history of taxonomy is reminiscent of Benoit Mandelbrot's paradox of the coastline. "When a bay or peninsula noticed on a map scaled to 1/100,000 is reexamined on a map at 1/10,000, subbays and subpeninsulas become visible," he wrote in his 1982 book *The Fractal Geometry of Nature*. "On a 1/1,000 scale map, sub-subbays and sub-subpeninsulas appear, and so forth. Each adds to the measured length." The length of the coastline, in other words, depends on the length of the ruler. So it is with life on Earth. As taxonomists moved beyond their sole reliance on physical features, drawing in statistical, ecological, reproductive, and geographic data, their view of the divisions among creatures grew finer, more nuanced. This was particularly true of genetic technologies, which have helped taxonomists reorganize whole sections of the tree of life and have revealed many new species, often physically indistinguishable from known species, at least to the human eye, but as genetically distinct as, say, a human and a bonobo. In recent decades, scientists have described "morphologically cryptic" bats, a kiwi bird, cicadas, butterflies, and pine trees, among many others. But these technologies, capable of revealing even the most minor variations, also threaten to drive what taxonomists have called "taxonomic inflation," the proliferation of spurious species descriptions based on flimsy DNA evidence; Hoser's critics place many of the creatures he has described into this category. Even the finest of rulers offers no solution to the puzzle that has bedeviled taxonomists ever since Linnaeus: While a species can be widely agreed upon, supported convincingly by data, it can't be proved. Where to draw the lines between species, genera, families, and so on remains subjective, a reflection of the individual taxonomist's understanding, a matter of philosophy.

This underlying uncertainty extends to the scientific debate over the number of species on Earth, the mysterious denominator in any question of extinction rates. Some estimates put the number at around two million, not counting single-celled organisms. This would mean Adam's task is three-quarters done. But most estimates are much higher. Some of the most generous numbers come from a 2017 paper, in which a group of researchers estimated that Earth is home to between two hundred million and nearly six billion species. The vast range of this estimate depends on a series of assumptions, said evolutionary biologist John Wiens, the paper's senior author. The majority of species on Earth, he said, are arthropods (a group that includes insects, lobsters, and spiders), and the mites, nematodes, protists (multicellular organisms that are not animals, plants, or fungi), and bacteria that inhabit those arthropods. "We tried to get everything," he said, "and it was under the assumption that what's living inside of insects will be the main drivers of biodiversity." The low end of the estimate comes from an assumption that each obviously distinct species of arthropod hides no cryptic species. The high end comes from an assumption that each distinct arthropod hides five cryptic species, a number the team arrived at based on previous studies of the ratio of morphologically distinct to cryptic arthropod species. Six billion species: Such diversity is godlike, awe-inspiring and incomprehensible, ungraspable, a matter of faith. It is, as Mandelbrot wrote of coastlines, "an elusive notion that slips between the fingers of one who wants to grasp it."

I was particularly struck by something I learned during a series of conversations with paleontologists. Over the past 3.7 billion years or so, the diversity of life on Earth has repeatedly expanded via the process of evolution, then contracted during catastrophic moments of mass extinction, then expanded again. The flowering plants, which appeared roughly 100 million years ago and diversified rapidly, drove parallel explosions in the diversity of insects and fungi. Aside from a mass-extinction event 65 million years ago, which killed off the nonavian dinosaurs, along with many other creatures, this expansion of diversity has continued mostly unabated. While the fragmentary nature of the fossil record and our incomplete knowledge of biodiversity in the present make comparisons difficult, the paleontologists I spoke with said there is general agreement that life on Earth is more diverse than it has ever been. This seems to me like a devilish paradox—that, at the moment of life's most magnificent bloom, there should appear a creature able not only to recognize, document, and name it, but also to bring about its destruction.

Facing a race against time and still uncertain about such basic details as how much remains to be done, some taxonomists have suggested ways of speeding up their collective task. Others predict that taxonomy as we know it will eventually be replaced by something wholly new. They include Richard Wells, gifted with the power of foresight. "The Linnaean system is what we follow. However, it is crap," he said. "It's a step on a stairway we can't even see the end of." Soon, human knowledge will undergo a sudden advance, he said, "and it's going to make every name absolutely irrelevant. All the glory that they attached to the naming of animals will just disappear in an instant." He could perceive the outlines of this future taxonomy but saw little point in bringing it into existence himself. "There's no point," he said. "All they give me is shit and arguments, and I don't really care, because life doesn't give a shit. Life just moves on. It doesn't give a shit whether Raymond Hoser has named it or not, or I named it. It doesn't matter. It's not about the people that name it. It's as transient as the wind blowing a fart away. It's just nothing."

~

Back in the parking garage at the mall, I helped Hoser pile the newspapers on top of the reptile crates. The Nissan Micra was now completely full. We raced from the lot onto side streets, arterials, boulevards, esplanades, promenades, and finally, his driveway. The Hosers live on a corner lot in Park Orchards, a spacious, wooded suburb northeast of Melbourne. The house is single-story brick, with a large, fenced, well-kept yard and a tall gum tree by the driveway. There are several outbuildings, including a large green shop made of corrugated steel. We carried the newspapers and reptiles from the Nissan Micra into the shop. Hoser was shirtless again and he was on the phone trying to reason with a man whose lace monitor was missing and possibly dangerous. "No, no, no, no, no, cut the bullshit!" he said. "They're not venomous! No, no, no, there's beliefs and then there's science. . . . Well, whatever you've been told is wrong. . . . Yeah, well, people go to hospital for all sorts of shit, including lace monitors, but that is not from venom."

The shop had a concrete floor and a high ceiling. There was a utility sink in the corner, a rusty birdcage and orange table by the door, and a poster on the back wall that read "Live Deadly Snake Show" above a picture of a smiling girl draped with a snake. There were several large tubs with lights and burbling pumps—one for the turtles, one for Joshua, and one for a second crocodile that hadn't attended Eva's birthday party. There were three large plywood racks on casters with thirty-six cubbies each. Each cubby held a plastic tub roughly six inches high, fifteen inches wide, and twenty-four inches deep. The tubs had airholes around their top rim, and many of them were labeled with handwritten species names. When Hoser saw me looking at one of these labels, he broke away from his conversation with the owner of the missing lace monitor. "Don't worry about the labels on the cages," he said. "They're all wrong." He explained that he had misnamed them deliberately to mislead thieves and wildlife authorities.

The room had a sour ammonia tinge, the same as in the Nissan Micra, only stronger. "They just eat, shit, eat, shit," Hoser told me when he was wheeling his newspapers through the mall. And it was true. The snakes that he'd taken to the birthday party had shit copiously, and now it was all over them and their tubs. He carried his taipans in their small birthday-party box to the utility sink, along with the larger tub where they ordinarily lived. Each tub contained a shelter made from the bottom of a five-gallon bucket, usually salvaged from alongside Melbourne's roads, as well as a yogurt-container water bowl set into a concrete anti-tipping fixture that he cast himself. He rinsed the squirming taipans under the faucet, then held them to the side while he rinsed their tub and swabbed it clean with newspaper. He lined it with fresh newspaper, refreshed the water bowl, returned the taipans to the tub, closed the lid, then slid the tub back into its cubby.

Other reptile keepers I spoke with said the cubby system is typical. The pet-store terrarium display with a light and plants and rocks is fine for someone who keeps one or two animals, they said, but is entirely impractical for someone who keeps hundreds. Reptiles like the cubbies, they told me; so long as a reptile is fed, watered, and provided with a suitable temperature, it is content. Supposing otherwise—that the reptiles might want more room to stretch out, might want to experience novel settings and sensations, might want to make decisions for themselves—is an exercise in anthropomorphism, these keepers said. But it seems to me the same is true of imagining reptilian contentedness in what is clearly a system of human convenience. Obsession is nothing if not self-justifying.

I could feel my own obsession beginning to uncoil. For years it had wound itself around me, a story so strange and convoluted that I often struggled to make heads or tails of it. What had begun with my sudden interest in the unfinished taxonomic task and an oddly scented piece of redwood had at times swelled to a grand scale, into a Manichaean battle between the forces of chaos and order, concerning no less than the future of life on Earth. But other times, like when I watched Raymond Hoser swabbing snake shit with pilfered Chinese-language newspapers, the story shrank back down. In these moments it was an absurdist farce, a tale less of the scientific quest for understanding or other high-minded pursuits than of man's folly, of his desire for immortality, of the impermanence of his works.

I also couldn't help but notice how little snakes had to do with any of it. Subjects of myth, of fear, of repulsion, of attraction, at no point had they played anything but a tertiary role, mute props to the snake men. They remained, to me at least, inscrutable, more easily understood as mirrors back on people than as entities in their own right. On a couple occasions during my time with Hoser, he left me for a minute or two by myself in the shop. Each time I was struck by the quiet—alone with hundreds of animals, I could hardly hear them at all.

After Raymond Hoser finished scrubbing the tubs of the animals he had taken to Eva's birthday party, we went into the house. We were met at the door by Slop, an enormous sway-hipped Great Dane who ate raw chicken wings for dinner, whom Hoser calls "child" and "good person," and who is the namesake of *Sloppopus, Slopboiga, Slopptyphlops, Sloppvipera, Sloppyscincus, Sloppopina, Sloppossum, Sloppyscinciini, Sloppophryne,* and at least two dozen other creatures. Hoser's wife Shireen, namesake of *Shireenhoserus, Shireenhosersaurea, Shireengeckiini, Shireengecko, Shireenaranea,* and many others, was standing at the sink, doing dishes. Their younger daughter, Jacky, namesake of *Jackyhosergecko, Jackyhoserhylea, Jackyhosersaur, Jackypython,* and more, returned from the beach and disappeared into another part of the house. Hoser offered tea. He took his with milk, Whiskas brand, preferred by eight out of ten cats, and Hoser, who does not own a cat but says that the flavor of kitten milk is "really, really nice."

After tea, he led the way to a second outbuilding. The door opened into a dim carpeted room that had a desk with his computer against one wall and a countertop with a stove and sink along another. The rest of the room was filled with plywood racks. There was the same ammonia smell. "More fucking animals," he said. "Snakes. Snakes, snakes, snakes, snakes, snakes. Snake shit. You get to know the different smells. I can even tell from the smell what snake has shit. There's lots of shits in this room." I sat on a stool by the desk as Hoser began cleaning plastic tubs in the sink. "I'll show you here, I've got a black snake that shat," he said, holding the snake in one hand while cleaning its tub with the other.

Next up was a tiger snake. "This is a tiger snake," he said, holding it up. "See how much it shits? Shit, shit, shit, shit, shit, just shit everywhere."

Shut out by herpetologists, Hoser had in recent years begun making forays into other taxonomic corners, describing new fish, funnel-web spiders, a tree kangaroo, a rock wallaby, pygmy possums, ring-tailed possums, yellow-bellied gliders, and potoroos. In a 2022 paper published in the journal *Australian Mammalogy*, a group of researchers noted the increasing problem of mammalian taxonomic vandalism appearing in self-published journals. While carefully avoiding Hoser's name, they pointed to the herpetological boycott against him and announced they would begin a similar effort. "We do not consider certain names published outside the peer-reviewed literature to be part of the permanent scientific record," the researchers wrote, "and they will be ignored by the Australian Mammal Society."

Much of Hoser's time, as ever, is taken up with his various lawsuits. During the COVID-19 pandemic, he appeared frequently in Australian and international newspapers, as he had throughout his life, offering commentary on snake events, snake trends, and his own snake-catching exploits. "If you don't know how to handle a snake, one wrong move could turn it into a killing machine," he said to a reporter for the *Daily Mail Australia*. "They are deadly, you get bitten by a tiger snake and it's pretty intense. They're quite erratic," he told *9News*. This latter article was about a call Hoser received to collect a tiger snake wrapped around the pump at a Coles Hardware petrol station. The next day, *9News* ran another article, this time writing, "*9News* can reveal today that the Coles Express staff did not call for help and that the rescue appears to be merely a stunt."

"I don't plant snakes to generate business," Hoser told the *Herald Sun* soon after. "I have heard of others doing it but not me."

A Coles spokesperson told the *Daily Mail Australia*: "There have been no recent reports of snakes at any Coles Express sites in Victoria. Coles is assisting Victoria Police with their investigation."

Soon after, Hoser sued several media outlets and reporters that had published articles about the petrol-station situation, arguing in his filing that these entities, along with the anonymous people who left comments on their websites, had made a vast number of defamatory imputations, including "The plaintiff is known for butchering his snakes," "The plaintiff is an idiot," and "The plaintiff is a serial pest in the reptile world." The judge dismissed most of Hoser's allegations but allowed him to replead a few of them, *The Age* reported, including the alleged imputation that he was a "faker and a fraud." By the time of my visit to Melbourne in early 2023, he still hadn't refiled.

He was busy with other matters, including a legal duel with another snake man over the snake man's public use of the phrase "snake man." Then there was perhaps his most ambitious legal gambit yet: a lawsuit against seven members of the Turtle Taxonomy Working Group for their failure to adopt his turtle names. In a 133-page affidavit, he noted his unparalleled credentials; described his decades-long fight against corrupt police, judges, bureaucrats, and functionaries; and recounted the entire history of his battle with the Wüster Gang, culminating in its boycott of his names, which, he noted, was in open defiance of the International Code of Zoological Nomenclature. He argued that in their recent publication, *Turtles of the World*, the targets of his suit had defamed him by accusing him of "nomenclatural claim-jumping," "intellectual plagiarism and unconscionable preemptive scientific appropriation of others' detailed and careful scientific work," "unethical practices," and "unscientific and disruptive behaviours." He requested that the court order the respondents to cease distribution of *Turtles of the World*, to "strictly comply" with the International Code of Zoological Nomenclature, to publish his provided 250-word apology, which he stipulated should be in Arial font in black print on a white background, and to each pay him $200,000 in damages.

During one of our conversations about his litigation, I asked Hoser whether he had any concerns about legal costs and countersuits. He waved the question away. Because he wrote and filed all his legal documents himself, mostly after business hours, there was no up-front cost to him. Any countersuits would be futile because he had put all his assets in his daughters' names. "I'm a straw man," he said.

~

As he swabbed tub after tub, Hoser told me about his enemies' continued efforts to excise him from the public memory. While most of the world's herpetologists were officially ignoring his names, they agreed that some of the creatures he had described actually existed and needed names, so they had begun replacing Hoser's names with new ones. His *Funkisaurus* thus became *Broadleysaurus*; his *Katrinahosertyphlops* became *Malayotyphlops*; his *Maxhoserus* became *Indotyphlops*. "We . . . completely disregard Hoser's nomenclature due to his broad acts of taxonomic vandalism, which do not stand up to even the slightest level of scientific standard or scrutiny," wrote one group of researchers in a 2022 paper, in which they renamed several skink species that Hoser had previously named. As of this writing, researchers have replaced more than one hundred of his names. Meanwhile, references to Kaiser's 2013 paper proposing the boycott have begun appearing frequently in recent works on herpetological taxonomy as a kind of disclaimer, often using nearly identical language. "Concerning the usage of unscientifically and unethically erected taxon names we follow the recommendations of 'censuring taxonomic vandals', as proposed by Kaiser et al. (2013)," reads one typical example.

Describing these developments, Hoser's voice rose to a pitch of utter disbelief. Then, still clutching two eastern brown snakes whose tub he was in the middle of washing, he crossed the room, pulled *Snakes of Western Australia* from the shelf, and set it down on the desk beside me. I tried to subtly move my legs farther from the brown snakes, which were possibly the beneficiaries of Hoser's devenomization surgery, possibly not; as various court documents have noted, it is virtually impossible to tell just by looking. Hoser flipped through the book, then stopped and pointed to a label at the top of the page: Pilbara death adder, *Acanthophis wellsi* Hoser, 1998. I recognized it as the beginning of the story. Whatever gambits and schemes the Wüster Gang come up with, he said, "They'll never take that from me."

That is probably true. The Pilbara death adder is one of a handful of creatures Hoser named before 2000, the starting year for the list of names under Kaiser's ongoing blockade. It is widely agreed to be the name of a previously unnamed species that actually exists. What becomes of the rest of his names is less certain. Hoser thinks that, in time, after he and Wüster and Kaiser and O'Shea and all the rest are dead, the boycott against him will fall and taxonomists will resurrect his names. He will take his rightful place in history as one of the most prolific taxonomists ever. That's possible. Maybe, just as his taxonomy represents a kind of rebellion, some future rebel will use his names to stick it to the scientific establishment of their day.

But as I sat there in the dim, ammonia-scented room, it also seemed possible that the embargo would hold, calcifying into a permanent barrier, and that Hoser would be remembered less for the creatures he'd named and more for the imaginative and controversial ways he'd named them, for the international imbroglio he sparked, for sowing disorder within taxonomy, the most arcane and august of institutions. It seems to me that this would be its own kind of immortality.

Hoser swabbed snake shit from a hundred or so tubs, then we ate a dinner of roasted vegetables and salt-and-pepper squid, prepared by Shireen, who listened quietly as he continued to tell me about his fights with the Wüster Gang, occasionally rolling her eyes at his most strident remarks. In "Medusa and the Taxonomic Vandal," the Australian poet Amanda Joy pointed to their marriage as a possible impetus for his taxonomic efforts.

> *A man naming one Shireen*
> *after his wife is labelled*
> *a taxonomic vandal*

> *Only she heard him*
> *correctly when he said it was*
> *a means of apology*

Whether it was a way to make amends, or the symptom of an incurable case of the mihi-itch, or a desire to sow chaos, as his enemies allege, or simply a scientific vocation he cares deeply about, as he maintains, Hoser kept naming. In the two years after my visit to Australia, he published eighteen more issues of the *Australasian Journal of Herpetology*. More than a decade into a near-universal boycott of his work, he seemed to have shed any vestigial concerns he may have had about how others might interpret his work, naming with freedom and abandon. He called a giant salamander *Andrias another*, "in reflection of what it is," he wrote, "namely another species in the genus *Andrias*." He named a lizard *Albaelacerta nonukesplease*. He named other lizards *Questionthenarrative always*, *Questionthenarrative constantly*, *Questionthenarrative continuously*, *Questionthenarrative asarule*, *Scienceneedsevidence yes*, and *Scienceneedsevidence toprogress*. He named a blind snake *Aa aaaaaugh* and another one *Zzzzz zzzzz*. He named species *K. corruptpoliceorum*, *L. fasciststateorum*, *L. dishonestpoliceorum*, *S. murderingpoliceorum*, and *H. nswpolicearecrooks*. He named species *Scored one*, *Scored another*, *Scored yetanother*. He named species *S. exy*, *O. arukidding*, *V. aruserious*, *C. wereisdat*, *Oh kay*, *Oh yes*, *Oh sheet*, *Oh phuk*, *C. whatdafuk*, *C. fukdat*, *R. datstink*, *S. shittythingie*, *J. shitbomb*, *F. tasteslikesheet*, and *S. ivebeenshaton*.

In these papers, he frequently discussed the Wüster Gang, which had swelled in his estimation to include all of the 450-some scientists who had signed their names to the boycott against him, accusing its members of being "hard-core criminals," "self-professed pseudo-scientists," "sex offenders and thieves," of terrorism, of trying to rename the Gulf of Oman "the Gulf of Wolfgang Wüster." He moved fluidly from these accusations into discussions of the various animals' scale shapes, colors, distributions, and habits.

Poring over Hoser's descriptions, I could not say whether *Zzzzz zzzzz* and *Oh sheet* and the others represented actual groups of heretofore unrecognized creatures that were even now living out their lives somewhere on Earth, or whether they were lines applied in arbitrary ways, subdividing species whose members would recognize no such cleavages themselves, or whether they were entirely products of Hoser's imagination. All of it seemed to reside in the liminal space between reality and myth, knowledge and delusion. At the same time, the work was clearly taxonomy. It was nomenclature. It was naming, something that people have done ever since Adam, sometimes for practical reasons, other times for lofty ideals, still other times out of greed or pride or because they could not seem to help themselves. They labor together on a task that is without beginning or end, that is orderly and chaotic, serious and silly, noble and base, an act of continual creation: Out of *not-species*, *species*. The best metaphor for humanity's eternal quest to describe and name the diversity of life, I came to think, is not a tower or a branching tree. It is a great serpent with its tail in its mouth. It is a hoop snake.

NOTES

CHAPTER 1: ADAM'S TASK

1 **By Spirits reprobate**: Milton, *Paradise Lost*, ed. John Leonard (Penguin, 2000), 20.
2 **Snakes of Western Australia**: G. M. Storr et al., *Snakes of Western Australia*, 2nd ed. (Western Australian Museum, 2002).
2 **a title he has trademarked**: "Legal Statement," n.d., smuggled.com.
3 **even to the name, "taxonomy"**: A. P. de Candolle, *Théorie Élémentaire de la Botanique* (Paris: Deterville, 1813).
6 **every beast of the field**: Genesis 2:19–20 (King James Version).
6 **he'd named some 12,000 species**: "His Career and Legacy," Linnean Society of London.
6 **about 17,000 new species every year**: Mark J. Costello et al., "More Taxonomists Describing Significantly Fewer Species per Unit Effort May Indicate That Most Species Have Been Discovered," *Systematic Biology* 62, no. 4 (April 10, 2013): 616–24.
6 **suburban New Jersey**: Jeremy A. Feinberg et al., "Cryptic Diversity in Metropolis: Confirmation of a New Leopard Frog Species (Anura: Ranidae) from New York City and Surrounding Atlantic Coast Regions," *PLoS One* 9, no. 10 (October 29, 2014): 1–15.
6 **new whales**: Patricia E. Rosel et al., "A New Species of Baleen Whale (Balaenoptera) from the Gulf of Mexico, with a Review of Its Geographical Distribution," *Marine Mammal Science* 37, no. 2 (February 11, 2021): 577–610; Shiro Wada et al., "A Newly Discovered Species of Living Baleen Whale," *Nature* 426 (November 20, 2003): 278–81.
6 **a new orangutan**: Alexander Nater et al., "Morphometric, Behavioral, and Genomic Evidence for a New Orangutan Species," *Current Biology* 27, no. 22 (November 20, 2017): 3487–98.
6 **a new elephant**: Alfred L. Roca et al., "Genetic Evidence for Two Species of Elephant in Africa," *Science* 293, no. 5534 (August 24, 2001): 1473–77.
8 **"confound[ing] their language"**: Genesis 11:7 (King James Version).
9 **"the Raymond Hoser problem"**: Darren Naish, "Taxonomic Vandalism and the Raymond Hoser Problem," *Tetrapod Zoology* (blog), *Scientific American*, June 20, 2013.
9 **"A Few Bad Scientists"**: Benjamin Jones, "A Few Bad Scientists Are Threatening to Topple Taxonomy," *Smithsonian Magazine*, September 7, 2017.

9 **editorials about him:** Wolfgang Wüster et al., "Confronting Taxonomic Vandalism in Biology: Conscientious Community Self-Organization Can Preserve Nomenclatural Stability," *Biological Journal of the Linnaean Society* 133, no. 3 (July 2021): 645–70; Wolfgang Wüster et al., "Taxonomic Contributions in the 'Amateur' Literature: Comments on Recent Descriptions of New Genera and Species by Raymond Hoser," *Litteratura Serpentium* 21, no. 3 (January 2001): 69–91.

10 **"Ye shall not die":** Milton, *Paradise Lost*, 203–4.

10 **"They spend little time on love":** Diane Morgan, *Snakes in Myth, Magic, and History* (Bloomsbury, 2008), 31.

10 **"milking a cow is no simple task":** Clifford B. Moore, "America's Mythical Snakes," *Scientific Monthly* 68, no. 1 (January 1949): 52–58.

10 **Egyptians carved a hoop snake:** Marinus Anthony van der Sluijs and Anthony L. Peratt, "The *Ourobóros* as an Auroral Phenomenon," *Journal of Folklore Research* 46, no. 1 (January 2008): 3–41; W. R. Cooper, *The Serpent Myths of Ancient Egypt* (London: Robert Hardwicke, 1873), 63.

11 **"Everybody is interested in snakes":** Eric Worrell, *Song of the Snake* (Angus and Robertson, 1958), 190.

11 **it is present in 38 percent of women:** Arne Öhman and Susan Mineka, "The Malicious Serpent: Snakes as a Prototypical Stimulus for an Evolved Module of Fear," *Current Directions in Psychological Research* 12, no. 1 (February 2001): 5–9.

11 **Researchers have discovered that people:** Arne Öhman, "The Role of the Amygdala in Human Fear: Automatic Detection of Threat," *Psychoneuroendocrinology* 30, no. 10 (November 2005): 953–58.

11 **"Here at least we shall be free":** Milton, *Paradise Lost*, 9.

14 *Acanthophis wellsi* **Hoser, 1998:** In fact, this species appears in Storr's *Snakes of Western Australia* as *A. wellsei*, an emendation of Hoser's Latin that is used by some taxonomists. For the sake of simplicity and consistency, I will stick with Hoser's original spelling throughout.

CHAPTER 2: AUSTRALIA TERRA COGNITA

15 *I tell you zat I had no thought:* "An Explanation: By a Distinguished Foreigner, and Member of the Philosophical Institute," *Melbourne Punch*, April 1, 1858. http://nla.gov.au/nla.news-article171432822.

15 **"I have named it Point Hicks":** James Cook, *Captain Cook's Journal During His First Voyage Round the World* (London: Elliot Stock, 1893).

16 **creatures he'd snagged with his dip net:** Joseph Banks, *Journal of the Right Hon. Sir Joseph Banks*, ed. Sir Joseph D. Hooker (London: MacMillan, 1896), 259–62.

16 **occasional differences of detail:** Carol Kaesuk Yoon, *Naming Nature: The Clash Between Instinct and Science* (W. W. Norton, 2009), 119.

16 **consistency of reality:** Brent Berlin, *Ethnobiological Classification: Principles of Categorization of Plants and Animals in Traditional Societies* (Princeton University Press, 1992), 8–9.

17 **Vervet monkeys scream:** Yoon, *Naming Nature*, 174–76.

17 **use around six hundred taxonomic names:** Yoon, *Naming Nature*, 138–40.

17 ***Achillea follies duplicatopinnatis:*** Rob Dunn, *Every Living Thing: Man's Obsessive Quest to Catalog Life, from Nanobacteria to New Monkeys* (Harper, 2009), 24; Michael Ohl, *The Art of Naming*, trans. Elisabeth Lauffer (MIT Press, 2018), 47.

17 **cut through the jumble:** Carolus Linnaeus, *Systema Naturae: Facsimile of the First Edition* (B. De Graaf, 1964). Originally published 1735.

17 **a method of naming:** Carolus Linnaeus, *Species Plantarum* (Stockholm: Laurentii Salvii, 1753).

18 **gave Linnaeus the honorific:** Magnus Lidén and David Morrison, "Misunderstand-

ings and Misrepresentations About Linné's Alleged Family Motto," *Genealogical World of Phylogenetic Networks* (blog), May 21, 2018.

18 **"Let us build a city and a tower":** Genesis 11:4–9 (King James Version).

18 **who called the people "Indians":** Banks, *Journal*, 263–65.

19 **as the poet Oodgeroo Noonuccal recounted:** Oodgeroo Noonuccal, *Dreamtime: Aboriginal Stories* (Lothrop, Lee & Shepard Books, 1994). Originally published as *Stradbroke Dreamtime* (Angus & Robertson, 1972).

19 **"bestowed names":** George R. Stewart, *Names on the Globe* (Oxford University Press, 1975), 5–6.

20 **1,400 new species of plants and animals:** P. J. Brownsey, "The Banks and Solander Collections: A Benchmark for Understanding the New Zealand Flora," *Journal of the Royal Society of New Zealand* 42, no. 2 (May 28, 2012): 131–37.

20 **contend was a taxonomic mistake:** Robert Borofsky, "Cook, Lono, Obeyesekere, and Sahlins," *Current Anthropology* 38, no. 2 (April 1997): 255–82.

20 **"like those of the weeping willow":** Banks, *Journal*, 271.

21 **"Upwards of thirty years have now elapsed":** Adrian Franklin, *Animal Nation: The True Story of Animals and Australia* (University of New South Wales Press, 2006), 43.

21 **naturalists stuck close to the coast:** Paul Humphries, "Wilhelm Blandowski's Contribution to Ichthyology of the Murray–Darling Basin, Australia," in "William Blandowski and His Contribution to Nineteenth Century Science and Art in Australia," special issue, *Proceedings and Transactions of the Royal Society of Victoria* 121, no. 1 (2009): 90–108 (hereafter cited as "Blandowski and His Contribution").

21 **thirteenth and youngest child:** Thomas A. Darragh, "William Blandowski: A Frustrated Life," in "Blandowski and His Contribution," 11–60.

22 **"I have been considered . . . as a *rebel*":** Darragh, "William Blandowski," 19.

22 **"Suppose him standing over the fire":** Darragh, "William Blandowski," 37.

22 **"several drays and various recalcitrant horses":** Paul Humphries, "Blandowski Misses Out: Ichthyological Etiquette in 19th-Century Australia," *Endeavour* 27, no. 4 (December 2003): 160–65.

23 **"a keg half full of decaying reptiles":** John Kean, "Observing Mondellimin, or When Gerard Krefft 'Saved Once More the Honour of the Exploring Expedition,' " in "Blandowski and His Contribution," 109–28.

23 **"He seemed to see himself more as a heroic explorer":** Harry Allen, ed., *Australia: William Blandowski's Illustrated Encyclopaedia of Aboriginal Australia* (Aboriginal Studies Press, 2010), 6.

23 **"gazed into the unknown wilds":** William Blandowski, "Recent Discoveries in Natural History on the Lower Murray," *Transactions of the Philosophical Institute of Victoria* 2 (1858): 124–37.

23 **alone into the wilderness:** Patrick White, *Voss* (Random House Australia, 1957).

24 **"appears to have been received as a minor celebrity":** Kean, "Observing Mondellimin," 115.

24 **"undoubtedly the highlight of Blandowski's career":** Darragh, "William Blandowski," 40.

24 **a paper with descriptions of nineteen species of fish:** Blandowski, "Recent Discoveries," 124–37.

24 **noted that Blandowski had not been present:** Darragh, "William Blandowski," 42.

25 **a fifty-six-line poem that included the stanzas:** "An Explanation: By a Distinguished Foreigner."

25 **He returned to Germany:** Darragh, "William Blandowski," 46–47.

25 **enlisted an art student:** Hannelore Landsberg and Marie Landsberg, "Wilhelm von Blandowski's Inheritance in Berlin," in "Blandowski and His Contribution," 183.

25 **admitted to the Silesian Provincial Mental Asylum:** Darragh, "William Blandowski," 48.

26 **their scholarly lenses:** Harry Allen and Elizabeth A. Weldon, eds., "William Blandowski and His Contribution to Nineteenth Century Science and Art in Australia," special issue, *Proceedings and Transactions of the Royal Society of Victoria* 121, no. 1 (2009).

26 **"a masterstroke":** Humphries, "Wilhelm Blandowski's Contribution," 104.

26 **kept a collection of human skulls:** Matthew Fishburn, "The Field of Golgotha," *Meanjin Quarterly* (Fall 2017).

26 **thanked his Aboriginal collaborators:** Humphries, "Wilhelm Blandowski's Contribution," 104.

26 **as fellow outcasts:** Harry Allen, "Introduction: Looking Again at William Blandowski," in "Blandowski and His Contribution," 5–6.

27 **"Sadly they disappoint the naturalist":** Allen, "Looking Again at William Blandowski," 4.

27 **a total of £313:** Darragh, "William Blandowski," 40.

27 **almost entirely erased:** Humphries, "Wilhelm Blandowski's Contribution," 90.

27 **distinguished but brief:** Jenny Nancarrow, "Gerard Krefft: A Singular Man," in "Blandowski and His Contribution," 140–54.

27 **his 1869 book:** Gerard Krefft, *The Snakes of Australia* (Sydney: Thomas Richards, 1869).

27 **"Body and tail somewhat elongated":** F. M. Daudin, *Histoire naturelle génerale et particuliére des reptiles, tome cinquiéme* (Paris: F. Dufart, 1803), 287–96.

28 **"considered a highly poisonous species":** George Shaw and E. R. Nodder, *Naturalist's Miscellany*, vol. 13 (London, 1802).

28 **the animal appeared in Krefft's *The Snakes of Australia*:** Krefft, *Snakes of Australia*, 80.

28 **"the deaf adder that stoppeth her ear":** Psalms 58:4 (King James Version).

28 **in an 1827 article in *The Australian*:** "New South Wales—No. VII," *Sydney Australian*, February 21, 1827, http://nla.gov.au/nla.news-article37071378. Mentioned again in "Mr. Martin's Austral-Asia," *Sydney Colonist*, October 26, 1837, http://nla.gov.au/nla.news-article31719972.

28 **went their separate evolutionary ways:** Calculation made using Time Tree of Life; see explanatory note in the chapter 8 section of the notes, under the bold phrase "This comparative form."

28 **an apparent malapropian shift:** "The Death Adder," *South Australian* (Adelaide), October 31, 1845, http://nla.gov.au/nla.news-article71603153.

29 **stealing gold from the museum:** Nancarrow, "Gerard Krefft," 140–54.

CHAPTER 3: THE MIHI-ITCH

31 **His father-in-law and biographer later reported:** William Jardine, *Memoirs of Hugh Strickland* (London: John van Voorst, 1858), iv.

32 **improvements to the weathercock:** Jardine, *Memoirs of Hugh Strickland*, xix.

32 **on such topics as:** Jardine, *Memoirs of Hugh Strickland*, pt 2: iii–v.

32 **"war against names of exotic origin":** "Report of a Committee Appointed 'To Consider of the Rules by Which the Nomenclature of Zoology May Be Established on a Uniform and Permanent Basis,'" *Reports of the Eleventh Meeting of the British Association for the Advancement of Science* (London: John Murray, 1845), 9:105.

32 **Strickland complained in a letter:** Hugh E. Strickland, "On the Arbitrary Alteration of Established Terms in Natural History," *Magazine of Natural History* 8 (1835): 40.

33 **an especially egregious example:** SDW, "With Regard to the Scientific Name of the Coalhood," *Magazine of Natural History* 7 (1834): 593.

34 **"The lovers of confusion have been hard at work":** Hugh Edward [Edwin] Strickland,

"On the Inexpediency of Altering Established Terms in Natural History," *Magazine of Natural History*, n.s., vol. 1 (1837): 127.

34 **"a few general rules":** Hugh E. Strickland, "Rules for Zoological Nomenclature," *Magazine of Natural History*, n.s., vol. 1 (1837): 173.

35 **"couched in a spirit of prejudice":** Jardine, *Memoirs of Hugh Strickland*, pt. 2: cxciv.

35 **leading scientific society:** L. C. Rookmaaker, "The Early Endeavours by Hugh Edwin Strickland to Establish a Code for Zoological Nomenclature in 1842–1843," *Bulletin of Zoological Nomenclature* 68, no. 1 (March 2011): 29–40.

36 **Charles Darwin admitted:** Rookmaaker, "Early Endeavours."

36 **including one on the Mauritanian dodo:** H. E. Strickland and A. G. Melville, *The Dodo and Its Kindred* (London: Reeve, Benham, and Reeve, 1848), 5.

36 **stepped back to avoid a coal train:** Jardine, *Memoirs of Hugh Strickland*, cclviii–cclix.

36 **"nomenclature has not improved":** Rookmaaker, "Early Endeavours," 39.

37 **a detailed history of these events:** Richard V. Melville, *Towards Stability in the Names of Animals: A History of the International Commission on Zoological Nomenclature, 1895–1995* (International Trust for Zoological Nomenclature, 1995).

38 **covers the naming of animals:** International Commission on Zoological Nomenclature [ICZN], *International Code of Zoological Nomenclature*, 4th ed. (International Trust for Zoological Nomenclature, 1999).

38 **seven so-called principles:** ICZN, *International Code*, 124.

39 **collection of other silly names:** Mark Isaak, "Curiosities of Biological Nomenclature," Curious Taxonomy.

40 **names ending in *anus*:** John Wright, *The Naming of the Shrew: A Curious History of Latin Names* (Bloomsbury, 2014), 54.

41 **derived from the word *caffer*:** Gideon H. Smith and Estrela Figueiredo, "Proposal to Add a New Article 61.6 to Permanently and Retroactively Eliminate Epithets from the Root *caf[e]r-* or *caff[e]r-* from the Nomenclature of Algae, Fungi and Plants," *Taxon* 70, no. 6 (December 2021): 1395–96.

41 **"derogatory or insulting to a person":** Timothy A. Hammer and Kevin R. Thiele, "Proposals to Amend Articles 51 and 56 and Division III, to Allow the Rejection of Culturally Offensive and Inappropriate Names," *Taxon* 70, no. 6 (December 2021): 1392–94.

41 **set aside the principle of priority's usual start date:** Len Norman Gillman and Shane Donald Wright, "Restoring Indigenous Names in Taxonomy," *Communications Biology* 3, no. 609 (October 23, 2020): 1–3.

41 **end the practice of naming species after people:** Patricia Guedes et al., "Eponyms Have No Place in 21st-Century Biological Nomenclature," *Nature Ecology & Evolution* 7 (March 13, 2023): 1157–60.

41 **Some raised practical concerns:** Pedro Jiménez-Mejías et al., "Protecting Stable Biological Nomenclatural Systems Enables Universal Communication: A Collective International Appeal," *BioScience* 74, no. 7 (July 2024): 467–72; Lou Jost et al., "Eponyms Are Important Tools for Biologists in the Global South," *Nature Ecology & Evolution* 7 (June 19, 2023): 1164–65; Sergei L. Mosyakin, "Eponyms in Biological Nomenclature and the Slippery Slope and Pandora's Box Arguments," *Ukrainian Botanical Journal* 80, no. 5 (October 28, 2023): 381–85; Rohan Pethiyagoda, "Policing the Scientific Lexicon: The New Colonialism?," *Megataxa* 10, no. 1 (June 13, 2023): 20–25.

42 **seemed impatient with the question:** Charles Darwin, *The Voyage of the Beagle and The Origin of Species* (Alfred A. Knopf, 2003), 578–79.

43 **dozens of ways of defining "species":** Sean Stankowski and Mark Ravinet, "Quantifying the Use of Species Concepts," *Current Biology* 31, no. 9 (May 10, 2021): 428–29.

43 **"I could have thus embraced any profession":** C. S. Rafinesque, *A Life of Travels and Researches in North America and South America* (Philadelphia: F. Turner, 1836), 10.

43 **5,500-line epic poem:** Constantine Samuel Rafinesque, *The World, or Instability: A Poem* (Philadelphia: J. Dobson, 1836), 65.

43 **He immediately spotted a plant that was new to him:** Rafinesque, *Life of Travels and Researches*, 14.

43 **cancer—induced, one medical historian proposed:** Charles T. Ambrose, "The Curious Death of Constantine Samuel Rafinesque (1783–1840): The Case for the Maidenhair Fern," *Journal of Medical Biography* 18 (August 2010): 165–73.

44 **he had described and named 6,700 new species:** Daniel Mosquin, "Constantine Rafinesque, a Flawed Genius," *Arnoldia* 70, no. 1 (July 2012): 2–10.

44 **"a complete monomania":** T. J. Fitzpatrick, *Rafinesque: A Sketch of His Life with Bibliography* (Historical Department of Iowa, 1911), 47.

44 **"the unending series of problems . . . raised by Rafinesque's work":** Mosquin, "Constantine Rafinesque."

44 **scientists now accept about fifty of Rafinesque's genera:** Mosquin, "Constantine Rafinesque."

44 **"a most disinterested benevolent man":** Neal L. Evenhuis, "The Life and Work of Francis Walker (1809–1874)," *Fly Times* 2 (September 27, 2018): 1–101.

45 **"More than twenty years too late":** "Obituary: Francis Walker," *Entomologist's Monthly Magazine* 11–12, nos. 121–144 (November 1874): 140–41.

45 **"He worked in a purely mechanical fashion":** Evenhuis, "Francis Walker (1809–1874)," 28.

45 **"eleven synonyms of the well-known *Eutachina rustica*":** Evenhuis, "Francis Walker (1809–1874)," 28.

45 **"Flies, you who have always given me my most cherished delights":** Neal L. Evenhuis et al., "Nomenclatural Studies Toward a World List of Diptera Genus-Group Names. Part I: André-Jean-Baptiste Robineau-Desvoidy," *Zootaxa* 2373 (February 26, 2010): 6.

46 **"I am unequal to the task":** Evenhuis, "Francis Walker (1809–1874)," 30.

46 **infected with *Mihisucht*:** Neal L. Evenhuis, "The 'Mihi-Itch'—a Brief History," *Zootaxa* 1890 (October 1, 2008): 59–68.

CHAPTER 4: SNAKE OIL

49 **"Then he rushed to the museum":** Andrew Barton "Banjo" Paterson, "Johnson's Antidote," in *The Man from Snowy River* (Sydney: Angus and Robertson, 1895).

49 **"the celebrated snake charmer":** "Snake Charmer in Trouble," *Launceston Examiner*, April 19, 1855, 2, http://nla.gov.au/nla.news-article36292363.

49 **"the notorious snake-charmer":** "Shoplifting," *Colonial Times*, June 1, 1854, 2, http://nla.gov.au/nla.news-article8776148.

49 **"the stout able man":** "The Snake Charmer Bitten," *Courier*, February 17, 1854, 2, http://nla.gov.au/nla.news-article2241897.

49 **"bilious-looking":** "Snake Charming," *Courier*, February 10, 1854, 2, https://trove.nla.gov.au/newspaper/article/2239174.

49 **"professor":** "Professor Charles Underwood's Antidote for the Bites of Snakes," *The Argus*, March 8, 1860, http://nla.gov.au/nla.news-article5678476.

49 **learned his tricks in North America:** "Snake Charmer," *Adelaide Times*, April 23, 1849, 3, http://nla.gov.au/nla.news-article206980565.

49 **from Amazonian shamans:** "Antidote to Snake Bites: Interesting and Dangerous Experiment," *Melbourne Age*, April 29, 1859, http://nla.gov.au/nla.news-article154838146.

49 **record of his birth:** I found a record of Underwood's birth through Ancestry.com, at the following URL: https://www.ancestry.co.uk/imageviewer/collections/1624/images/31547_212661-00366?rc=&queryId=f9fd07c7-5eee-478f-98dd-bff24cea8ad7&usePUB=true&_phsrc=auQ3&_phstart=successSource&pId=10427023.

49 **highway robbery:** England & Wales, Criminal Registers, 1791–1892. https://www. ancestry.co.uk/search/collections/1590/records/1262046?tid=&pid=&queryId=ee5 2da63-6524-4e71-83d7-d3572be45dfb&_phsrc=auQ7&_phstart=successSource.

50 **aboard the barque *Norfolk*:** "Norfolk Voyage: 20th May 1829–27th Aug 1829," Convict Records, https://convictrecords.com.au/ships/norfolk/voyages/242.

50 **Queensland newspaper wondered:** "Cure of Snakebites, &c.," *Moreton Bay Courier*, December 29, 1849, 3, http://nla.gov.au/nla.news-article3714244.

50 **convicted of forging a check:** *Australasian Chronicle*, February 11, 1841, 3, http://nla. gov.au/nla.news-article31730824.

50 **it was as a snake man:** "Snakes," *Launceston Examiner*, April 4, 1849, 5, http://nla.gov. au/nla.news-article36257080.

50 **arranged a snakebite exhibition:** "Snake Charmer," *Hobart Town Advertiser*, April 10, 1849, http://nla.gov.au/nla.news-article703637.

51 **"only after you have slept with me":** Herodotus, *The Histories*, trans. Tom Holland (Viking, 2014), 266.

51 **"the limbs of women and their ways":** Dante Alighieri, *Inferno*, trans. Allen Mandelbaum (Alfred. A. Knopf, 1995), 95.

51 **Drops of Medusa's blood:** Ovid, *Metamorphoses*, trans. David Raeburn (Penguin Classics, 2004), 161.

51 **Cato led a Republican army:** Lucan, *Pharsalia*, trans. H. T. Riley (George Bell and Sons, 1909), 367–72.

51 **"a fine boy":** *Sydney Gazette and New South Wales Advertiser*, November 20, 1808, 1, http://nla.gov.au/nla.news-article627627.

51 **"a fine cow":** *Sydney Gazette and New South Wales Advertiser*, May 28, 1809, 1, http:// nla.gov.au/nla.news-article627748.

51 **"a fine ewe":** *Sydney Gazette and New South Wales Advertiser*, December 2, 1804, 3, http://nla.gov.au/nla.news-article626530.

51 **teeth turned black:** *Sydney Gazette and New South Wales Advertiser*, December 23, 1815, 2, http://nla.gov.au/nla.news-article629251.

52 **"instantaneously a mass of putrefaction":** *Sydney Morning Herald*, May 21, 1845, 3, http://nla.gov.au/nla.news-article12879665.

52 **remedies for snakebite:** John Cann, *Snakes Alive! Snake Experts and Antidote Sellers of Australia*, revised 2nd ed. (E.C.O., 2001), 32–49; Peter Mirtschin, "The Pioneers of Venom Production for Australian Antivenoms," *Toxicon* 48 (2006): 899–918; Ramona Morris and Desmond Morris, *Men and Snakes* (McGraw-Hill, 1965), 87–95.

52 **"28½ per cent died after the application":** "Antidote Against the Bite of Snakes," *Geelong Advertiser*, March 4, 1850, 2, http://nla.gov.au/nla.news-article93139612.

52 **advertising his antidote:** "Antidote for the Bite of Snakes and Other Venomous Reptiles," *Hobarton Guardian*, October 1, 1853, 1, http://nla.gov.au/nla.news-article172860451.

53 **stealing five gallons of rum:** "A 'Dealer' in Spiritous Liquors," *Tasmanian Colonist*, May 3, 1852, 2, http://nla.gov.au/nla.news-article226522884.

53 **"He places the reptile's head in his mouth":** "Snake Charming."

53 **"he was under the influence of liquor":** "Snake Charmer Bitten."

53 **threatened him with a bag of snakes:** "Snake Charmer in Trouble."

53 **Mr. Millhouse's Manufactory:** "Antidote for the Bite of Snakes and Other Venomous Reptiles," *Colonial Times*, April 16, 1857, 3, http://nla.gov.au/nla.news-article8785254.

53 **the death of George Henwood:** "Snake Charming: The Late Death from Snake Charming," *Courier*, December 24, 1858, 3, http://nla.gov.au/nla.news-article2465884.

53 **"necessary to warn the public":** "Snake Bites," *South Australian Register*, April 14, 1859, 3, http://nla.gov.au/nla.news-article49902077.

53 **"persons are politely requested":** *Geelong Advertiser*, April 20, 1859, 2, http://nla.gov. au/nla.news-article150077516.

54 **"no other than the common fern!":** "Underwood's Snake Antidote Discovered," *Courier*, May 5, 1859, 2, http://nla.gov.au/nla.news-article2470428.

54 **"the small sum of one penny per head":** "Antidote to Snake Bites," *The Argus*, May 9, 1859, 5, http://nla.gov.au/nla.news-article5680831.

54 **roughly 3,000 pounds:** *Geelong Advertiser*, May 10, 1859, 2, http://nla.gov.au/nla.news-article150078110.

54 **"could not be *the* Underwood":** "An Evening with the Snakes," *Bendigo Advertiser*, October 31, 1859, 2, http://nla.gov.au/nla.news-article87993189.

55 **Joseph Beaumont Shiers:** "Surrey 4, 18 August 1842," Libraries Tasmania, Convict Records, https://libraries.tas.gov.au/Digital/CON28-1-1/CON28-1-1_170.

55 **the file on his 1841 conviction:** "CON33-1-27 Image 198," Libraries Tasmania, Convict Records, https://libraries.tas.gov.au/Digital/CON33-1-27/CON33-1-27p198.

55 **Joseph Shiers was transformed:** "Convict Department," *Hobart Town Advertiser*, March 13, 1846, 4, https://trove.nla.gov.au/newspaper/article/264515607.

55 **fighting in bars:** "Taking Satisfaction," *Launceston Examiner*, September 28, 1850, 4, http://nla.gov.au/nla.news-article36267109.

55 **stealing a watch:** "Court of Quarter Sessions," *Colonial Times*, April 4, 1855, 2, http://nla.gov.au/nla.news-article87798.

55 **deserting a whaling ship:** "Deserting a Whaler," *Colonial Times*, July 16, 1856, 3, http://nla.gov.au/nla.news-article8784475.

55 **beating his wife:** "Assault of a Husband upon His Wife," *Courier*, August 9, 1854, 2, http://nla.gov.au/nla.news-article2242931.

55 **in connection with snakes:** "Antidote to Snake Bites," *Age*, August 15, 1859, 1, https://trove.nla.gov.au/newspaper/article/154828553.

56 **"a motley assemblage":** "Snake Charming," *Launceston Examiner*, January 26, 1860, 3, http://nla.gov.au/nla.news-article38998955.

56 **another advertisement appeared:** "Antidote for Snake-Bites," *The Argus*, April 30, 1860, 8, http://nla.gov.au/nla.news-article5681505.

57 **both men appeared to be drunk:** "Underwood and His Snakes," *Geelong Advertiser*, May 14, 1860, 2, http://nla.gov.au/nla.news-article148790868.

57 **the report on the inquest:** "Inquest on Underwood, the Discoverer of an Antidote for Snake Bites," *Age*, December 31, 1861, 5, http://nla.gov.au/nla.news-article154847851.

57 **a description of his investigations:** "Experiments on the Poison of the Cobra-de-Capella," *The Argus*, April 26, 1867, 6, http://nla.gov.au/nla.news-article5765221.

57 **like that of cholera:** "Snake Bite and Cholera," *Empire*, June 27, 1867, 3, http://nla.gov.au/nla.news-article60841814.

58 **"it will alter the whole system of pathology":** "The News of the Day," *Age*, September 10, 1867, 4, http://nla.gov.au/nla.news-article185504609.

58 **publicly invited Shires:** *Bacchus Marsh Express*, November 30, 1867, 5, http://nla.gov.au/nla.news-article88373832.

58 **"I caused him to be bitten":** "Shires's Antidote for Snake Bite," *The Argus*, December 5, 1867, 6, http://nla.gov.au/nla.news-article5785195.

58 **"As regards the value of the antidote":** "Professor Halford's Experiments with Shires' Antidote," *The Argus*, December 20, 1867, 6, http://nla.gov.au/nla.news-article5786396.

58 **manslaughter or murder charges:** "Snakes and Their Bites," *Leader*, December 7, 1867, 17, http://nla.gov.au/nla.news-article196636973.

58 **ordinary white blood cells:** "White Corpuscles of Blood, or Germ Cells," *Australasian*, November 8, 1867, 6, http://nla.gov.au/nla.news-article8850872.

58 **a mocking letter:** "Shires' Antidote for Snake Bites," *Melbourne Punch*, January 30, 1868, 6, http://nla.gov.au/nla.news-article174536962.

58 **"there is no deception about this man":** "Professor Halford's Snake Poison Experiments," January 23, 1868, 6, http://nla.gov.au/nla.news-article5789331.

59 **"saved him from being accessory to my death":** "Shires' Antidote," *The Argus*, January 17, 1868, 7, http://nla.gov.au/nla.news-article5788839.

59 **"Why cannot I parade my cures"**: "Again Shires' Antidote," *Ovens and Murray Advertiser*, January 14, 1868, 4, http://nla.gov.au/nla.news-article197438498.

59 **bite a police magistrate:** "Fatal Experiment with a Snake," *Age*, May 4, 1868, 5.

59 **Shires was eventually acquitted:** "Victoria," *Perth Gazette and West Australian Times*, July 10, 1868, 3, http://nla.gov.au/nla.news-article3749637.

59 **"This mode of treatment"**: "Professor Halford's Experiment on Snake-Poisoning," *The Argus*, November 7, 1868, 6, http://nla.gov.au/nla.news-article5831744.

59 **Shires's proprietary antidote:** *Bacchus Marsh Express*, November 30, 1867, 5, http://nla.gov.au/nla.news-article88373832.

59 **snakebite victims miraculously cured:** "Successful Treatment of Snake-Bite," *Ovens and Murray Advertiser*, December 5, 1868, 1, http://nla.gov.au/nla.news -article198055338.

59 **"Important to Squatters"**: "Snake Bites!! Snake Bites!!," *Sydney Morning Herald*, March 8, 1869, 6, http://nla.gov.au/nla.news-article13184604.

59 **known as the Halford syringe:** "A Chapter on Serpents," *Border Watch* (Beechworth, Vic.), July 17, 1880, 4; "Snake Cure Extraordinary," *Newcastle (N.S.W.) Chronicle*, November 28, 1868, 2, http://nla.gov.au/nla.news-article77588728.

59 **died despite receiving ammonia:** "The Late Death of a Woman from Snake-Bite," *Sydney Empire*, February 27, 1869, http://nla.gov.au/nla.news-article60832368.

59 **ought not inject themselves with ammonia:** "Snake Bite," *Sydney Morning Herald*, March 12, 1869, http://nla.gov.au/nla.news-article13180866.

59 **among those who wrote letters:** "To the Editor of the Herald," *Sydney Morning Herald*, March 17, 1869, http://nla.gov.au/nla.news-article13183288.

60 **the dogs all died:** "The Snake Experiments," *McIvor Times and Rodney Advertiser* (Vic.), April 13, 1876, http://nla.gov.au/nla.news-article91836264.

60 **"the speediest and most certain method"**: "The Intra-Venous Injection of Ammonia," *The Argus*, April 24, 1876, http://nla.gov.au/nla.news-article7437544.

60 **"other hands than his own"**: "An Interesting Identity," *Bendigo (Vic.) Independent*, July 4, 1891, http://nla.gov.au/nla.news-article169573446.

60 **"tree falling on him"**: "Deaths," *Melbourne Weekly Times*, March 26, 1892, http://nla.gov.au/nla.news-article220477041.

60 **"When the poor fellow gets well"**: "A Snake Showman and His Pets," *Bendigo Independent*, May 3, 1893, https://trove.nla.gov.au/newspaper/article/174255392; "Death from Snakebite," *Riverine Herald* (Echuca, Vic.), May 5, 1893, https://trove.nla.gov.au/newspaper/article/114972118.

61 **"I will be all right"**: "Snake Charmer's End," *Launceston (Tas.) Examiner*, May 5, 1903, http://nla.gov.au/nla.news-article35544891.

61 **set up among the jugglers:** John Cann and Jimmy Thomson, *The Last Snake Man: The Remarkable True-Life Story of an Aussie Legend and a Century of Snake Shows* (Allen and Unwin, 2018), 3–11.

61 **On See's first turn:** Cann, *Snakes Alive*, 106.

61 **"the second time within a week"**: "Deaths of Two Snake Charmers," *Brisbane Courier*, December 17, 1913, http://nla.gov.au/nla.news-article19919536.

61 **met the same fate:** "Professor" Fox, *Perth Truth*, March 7, 1914, http://nla.gov.au/nla.news-article207425235.

61 **Albarez was bitten:** Crag Wartell [Eric Worrell], "Snake-Men Play at Death," *Man Junior*, December 1952, Eric Worrell Archive.

61 **"merely a scratch"**: "Fangless Reptiles May Kill," *Sydney Sun*, March 21, 1917, http://nla.gov.au/nla.news-article221960261.

61 **"KING of all OINTMENTS"**: "That Man Gray," *Sydney Sun*, March 14, 1917, http://nla.gov.au/nla.news-article221964513.

61 **"90 per cent. more medicinal power"**: "A Remarkable Success," *Grafton (N.S.W.) Daily Examiner*, June 14, 1918, http://nla.gov.au/nla.news-article195138782.

61 **Cleopatra Caton was bitten:** Cann, *Snakes Alive*, 74.

62 **"tenth bite in South Africa"**: "Fatal Snake Bite," *Sydney Morning Herald*, September 2, 1921, http://nla.gov.au/nla.news-article28088130.

62 **"The green mamba wins"**: "The Green Mamba Wins," *Sydney Sunday Times*, August 28, 1921, http://nla.gov.au/nla.news-article123246802.

62 **He died, too:** "Snake 'Charmer's' Death," *Sydney Sun*, January 12, 1922, http://nla.gov.au/nla.news-article225228029.

62 **"confidence in a remedy of his own"**: "Snake-Charmer's Death," *Geelong (Vic.) Advertiser*, April 27, 1922, http://nla.gov.au/nla.news-article165979030.

63 **"no cause for alarm"**: "Snakebite Remedies Fail," *Mackay (Qld.) Daily Mercury*, June 24, 1927, http://nla.gov.au/nla.news-article172660666.

63 **Dot "Cleopatra" Vane was bitten:** "Snake Charmer's Death," *Kalgoorie (W. Aust.) Miner*, January 25, 1928, http://nla.gov.au/nla.news-article94354585.

63 **"not drunk at the time"**: "Snake-Charming," *West Australian* (Perth), March 6, 1929, http://nla.gov.au/nla.news-article32263686.

63 **"expressed extraordinary confidence"**: "Bitten by Snake," *Burnie (Tas.) Advocate*, April 9, 1931, http://nla.gov.au/nla.news-article67711596.

63 **Cecil O'Sullivan was bitten:** "Antidote Was Stolen," *Rockhampton (Qld.) Evening News*, November 25, 1932, http://nla.gov.au/nla.news-article201403450.

63 **snake man John Greaves:** "Fatal Bite," *Hobart (Tas.) Mercury*, November 5, 1932, http://nla.gov.au/nla.news-article24721364.

63 **"Milo the Snake Man"**: "Milo, 'the Snake Man,' Dead," *Albury (N.S.W.) Border Morning Mail*, December 24, 1934, http://nla.gov.au/nla.news-article271713873.

63 **"Captain Gus Leighton"**: Cann, *Snakes Alive*, 132.

64 **his experiments with rattlesnake venom:** Henry Sewall, "Experiments on the Preventive Inoculation of Rattlesnake Venom," *Journal of Physiology* 8 (August 1887): 203–10.

64 **immunize horses against cobra venom:** Albert Calmette, *Venoms, Venomous Animals and Antivenomous Serum-Therapeutics*, trans. E. Austen (John Bale, Sons & Danielsson, 1908).

64 **Calmette's antivenom offered no protection:** Peter J. Featherstone and Christine M. Ball, "The Development of Snake Antivenoms in Australia," *Anaesthesia and Intensive Care* 50, no. 5 (September 8, 2022): 342–44.

64 **injected a horse with tiger snake venom:** Frank Tidswell, "A Preliminary Note on the Serum-Therapy of Snake-Bite," *Indian Medical Gazette* (January, 1903): 33–35.

CHAPTER 5: SONG OF THE SNAKE

67 **"reptiles were undesirable"**: Kevin Markwell, "Relating to Reptiles: An Autoethnographic Account of Animal-Leisure Relationships," *Leisure Studies* 38, no. 3 (November 13, 2018): 341–52.

67 **"I idolised him"**: Markwell, "Relating to Reptiles."

67 **"Snake Bite Need Not Be Fatal"**: Eric Worrell, "Snake Bite Need Not Be Fatal," *Outdoors and Fishing*, May 1948, 186–87, Eric Worrell Archive.

67 **"Be Your Own Taxidermist"**: Belvedere [Eric Worrell], "Be Your Own Taxidermist," *Outdoors and Fishing*, February 1949, 386–87, Eric Worrell Archive.

67 **Sap-Engro, Karliboodi, and Belvedere:** Sap-Engro [Eric Worrell], "Cry Poacher," *Outdoors and Fishing*, November 1954, 26–29, 60, Eric Worrell Archive; Karliboodi [Eric Worrell], "Art and the Aborigine," *Outdoors and Fishing*, October 1948, 98–99, Eric Worrell Archive; Belvedere, "Be Your Own Taxidermist."

68 **"only good snake is a dead snake"**: Markwell, "Relating to Reptiles."

68 **beautiful baby contest:** Kevin Markwell and Nancy Cushing, *Snake-Bitten: Eric Worrell and the Australian Reptile Park* (University of New South Wales Press, 2010), 4.

69 **Kevin Markwell owned a book:** Roy Norry, *Australian Snake Man: The Story of Eric Worrell* (Thomas Nelson, 1966).

69 **a little boy standing beside the pit:** Cann, *Snakes Alive*, 139.

69 **there Cann met Snakey George:** Cann, *Snakes Alive*, 126.

70 **When he was sixteen:** Cann, *Snakes Alive*, 106–11.

70 **"gave me a whip snake":** Worrell, *Song of the Snake*, 145.

70 **a spirit of acquiescence:** Markwell and Cushing, *Snake-Bitten*, 6.

70 **"Everybody has a dream":** Worrell, *Song of the Snake*, 144.

71 **a series of odd jobs:** Markwell and Cushing, *Snake-Bitten*, 8–11.

71 **"mysterious or evil-smelling":** Worrell, *Song of the Snake*, 3.

71 **"easily handled ferrets are essential":** Eric Worrell, "You Can Have Fun with a Ferret," *Outdoors and Fishing*, October 1948, 90–91, Eric Worrell Archive.

71 **Snakes "do not milk cows":** Eric Worrell, "I Like Snakes," *Cavalcade*, March 1949, 58–60, Eric Worrell Archive.

72 **"white-throated ta-ta lizards"** Worrell, *Song of the Snake*, 17, 6–7.

72 **"passion-vines drape in wild profusion":** Eric Worrell, "The Wet Comes to Darwin," *Wild Life*, November 1947, 401–4, Eric Worrell Archive.

72 **"there were no more crocodiles":** Worrell, *Song of the Snake*, 27.

72 **"a week-end's snake shooting":** Worrell, *Song of the Snake*, 105.

72 **ornithologist Dom Serventy:** Worrell, *Song of the Snake*, 203.

73 **"their doom is fixed":** Darwin, *Beagle and The Origin*, 453.

73 **order of nature was strangely perverted:** Thomas R. Dunlap, "Remaking the Land: The Acclimatization Movement and Anglo Ideas of Nature," *Journal of World History* 8, no. 2 (Fall 1997): 303–19.

73 **"animals indigenous to Australia":** Edward Wilson, "The Distribution of Animals," *The Argus*, December 28, 1858, http://nla.gov.au/nla.news-article7307070.

74 **"kill hundreds with a stick":** Eric C. Rolls, *They All Ran Wild: The Animals and Plants That Plague Australia* (Angus and Robertson, 1969), 41.

74 **"the freedom of the bush":** An Old Bushman [Horace William Wheelwright], *Bush Wanderings of a Naturalist* (London: Routledge, Warne & Routledge, 1861), ix.

75 **"to utterly destroy native animals":** Brett J. Stubbs, "From 'Useless Brutes' to National Treasures: A Century of Evolving Attitudes Towards Native Fauna in New South Wales, 1860s to 1960s," *Environment and History* 7, no. 1 (February 2001): 23–56.

75 **a source of pride and cultural identity:** Franklin, *Animal Nation*.

75 **"it became hateful to all":** Edward Topsell, *The History of Four-Footed Beasts and Serpents* (London: Printed by E. Cotes for G. Sawbridge, T. Williams, and T. Johnson, 1658), 605. Topsell's original *History of Serpents* was combined with *The History of Four-Footed Beasts* in this edition.

75 **"stowed in the cultural baggage":** Markwell and Cushing, *Snake-Bitten*, xiv.

75 **"to kill every snake you see":** Peter Hobbins, *Venomous Encounters: Snakes, Vivisection and Scientific Medicine in Colonial Australia* (Manchester University Press, 2017), 113.

75 **"for every snake's head":** Old Bushman, *Bush Wanderings of a Naturalist*, 205.

75 **"have been considerably reduced":** Krefft, *Snakes of Australia*, iii.

75 **"We can't blame the serpent":** Worrell, *Song of the Snake*, 94.

76 **"one in two million":** Worrell, *Song of the Snake*, 90.

76 **nearly 70 percent of snakebite deaths:** Ravikar Ralph et al., "The Timing Is Right to End Snakebite Deaths in South Asia," *BMJ* 364 (January 22, 2019): 1–6.

76 **106 Australian snake handlers:** Geoffrey K. Isbister and S. G. A. Brown, "Bites in Australian Snake Handlers—Australian Snakebite Project (ASP-15)," *QJM* 105, no. 11 (November 2012): 1089–95.

76 **"the four T's":** Joshua D. Jaramillo et al., "The 'T's' of Snakebite Injury in the USA: Fact or Fiction?," *Trauma Surgery & Acute Care Open* 4, no. 1 (October 30, 2019): 1–6.

76 **"The way to avoid snakebite":** Worrell, *Song of the Snake*, 89.

77 **"animals not guarded by law":** Worrell, *Song of the Snake*, 207.

77 **"reptile hunter"**: "Crocodiles Cause Lilyfield Alarm," *Sydney Sun*, December 18, 1946, http://nla.gov.au/nla.news-article229541802.

77 **"snake collector"**: "Native Cat Too Wild to Film," *Evening Advocate* (Innisfail, Qld.), July 25, 1951, http://nla.gov.au/nla.news-article212299670.

77 **"professional snake catcher"**: "The Show Must Go On," *Perth Daily News*, November 5, 1952, http://nla.gov.au/nla.news-article266045107.

77 **"an expert in reptiles"**: "He Caught a 'Croc,'" *Queensland Times*, February 10, 1948, http://nla.gov.au/nla.news-article125582690.

77 **"naturalist"**: "No Taipans in N.T., Naturalist Believes," *Sydney Daily Mirror*, October 13, 1951, http://nla.gov.au/nla.news-article275328973.

77 **"herpetologist"**: "Croc-Catcher's off on the Job Again," *Sydney Sun*, February 4, 1948, http://nla.gov.au/nla.news-article229038932.

77 **"snake man"**: "Arthur Polkinghorne's Sydney Diary," *Sydney Sun*, November 7, 1952, http://nla.gov.au/nla.news-article230735396.

77 **"Mr. Worrell claims to have been bitten"**: "Nine Pythons in Hotel," *Port Augusta (S. Aust.) Transcontinental*, October 31, 1947, http://nla.gov.au/nla.news-article168308875.

77 **"It's nothing to find a small snake"**: Jean Powe, "Reptile Hunter Back from Northern Territory with Grisly Collection for Zoos," *Australian Women's Weekly*, January 18, 1947, http://nla.gov.au/nla.news-article47506109.

77 **"snake handler"**: "Bride Will Hunt Crocs, Adders," *Sydney Sun*, July 17, 1948, http://nla.gov.au/nla.news-article229007648.

77 **"snake woman"**: "Snake Woman Finds a Mate," *Sydney Sun*, July 25, 1948, http://nla.gov.au/nla.news-article229004815.

77 **"Snakes! They don't worry me"**: "Bride Will Hunt Crocs."

77 **"definitely not slimy"**: "Snake Woman Finds a Mate."

78 **primarily a reptile house:** Markwell and Cushing, *Snake-Bitten*, 33.

78 **lost in the mail:** "Taipan Astray in Mail?," *Mackay (Qld.) Daily Mercury*, August 3, 1950, http://nla.gov.au/nla.news-article172306135.

78 **"prohibited as a postal consignment"**: "Snakes Must Not Be Sent by Mail," *Newcastle (N.S.W.) Morning Herald and Miners' Advocate*, August 3, 1950, http://nla.gov.au/nla.news-article135293422.

78 **regularly producing tiger snake antivenom:** Kenneth D. Winkel et al., "Twentieth Century Toxinology and Antivenom Development in Australia," *Toxicon* 48, no. 7 (December 1, 2006).

78 **"I'm not afraid of snakes"**: "Man Seizes Snake in Pit," *Sydney Daily Telegraph*, November 23, 1951, http://nla.gov.au/nla.news-article248703623.

79 **"Operation Crocodile"**: "Queensland: Operation Crocodile," *British Pathé*, 1964.

79 **"Snakes do not like milk"**: Eric Worrell, "Don't You Believe It," *Australian Country*, August 1956, 98, Eric Worrell Archive.

79 **"not particularly well-written"**: "Snakes," *Bulletin* 79, no. 4088 (June 18, 1958): 58–59.

79 **"wide of the mark"**: "Snake Yarns," *Pacific Islands Monthly*, June 1, 1958, 89, https://nla.gov.au:443/nla.obj-320274159.

79 **tourism a secondary concern:** Markwell and Cushing, *Snake-Bitten*, 78–80.

79 **"graceful and realistic"**: Markwell and Cushing, *Snake-Bitten*, 91.

80 **a small snake from Queensland:** Eric Worrell, "A New Elapine Snake from Queensland," *Proceedings of the Royal Zoological Society of New South Wales* 74 (1953–54): 41–43.

80 **"These are understandable errors"**: Kevin Markwell, "Song of the Snake Man: Exploring the Writings of Eric Worrell," Eric Worrell Archive.

81 **"he saw a hoop-snake"**: Eric Worrell, *Reptiles of Australia* (Angus and Robertson, 1963), xi.

81 **"the audience for whom it was intended"**: Arnold G. Kluge, review of *Reptiles of Australia*, by Eric Worrell, *Copeia* 1965, no. 1 (March 18, 1965): 118–19.

81 **"clearly the work of an amateur"**: Eric R. Pianka, review of *Australian Lizards*, by Robert Bustard, *Copeia* 1971, no. 4 (December 1, 1971): 764–66.

81 **two more taxonomic works**: Eric Worrell, "A New Elapine Generic Name," *Australian Reptile Park Records* 1 (April 30, 1963): 1–7, Eric Worrell Archive; Eric Worrell, "Two New Subspecies of the Elapine Genus *Notechis* from Bass Strait," *Australian Reptile Park Records* 2 (June 1, 1963): 1–11, Eric Worrell Archive.

83 **"the help of my many friends"**: Eric Worrell, *Australian Wildlife* (Angus and Robertson, 1966), dust jacket.

83 **the questions of one Mr. Humphries**: "Venomous Snakes," *Hansard, New South Wales Legislative Assembly Printed Questions and Answers* (August 12, 1970), 5077–78.

83 **"They refuse to eat and pine away"**: *Hansard, New South Wales Legislative Assembly*, 6122–23.

84 **selling at a loss**: Markwell and Cushing, *Snake-Bitten*, 138.

84 **"mice, fowl, pigeons, chicks"**: Markwell and Cushing, *Snake-Bitten*, 47.

84 **bald defiance of his vision**: Markwell and Cushing, *Snake-Bitten*, 123–25.

84 **His drunkenness increased notably**: Markwell and Cushing, *Snake-Bitten*, 121.

84 **a list of rare and dangerous reptiles**: Markwell and Cushing, *Snake-Bitten*, 145–46.

85 **Legislative discussion**: "National Parks and Wildlife (Amendment) Bill," *Hansard, Legislative Council* (September 22, 1971), 1297–99.

85 **announced new regulations**: "NSW to License Reptile Collectors," *Sydney Morning Herald*, January 24, 1974.

85 **"Keeping reptiles was who I was"**: Markwell, "Relating to Reptiles."

86 **"For he was outlawed"**: Howard Pyle, *The Merry Adventures of Robin Hood* (New York: Charles Scribner's Sons, 1883), 4.

86 **"unpopular ones"**: Eric Worrell, "The Unpopular Ones," in *The Great Extermination: A Guide to Anglo-Australian Cupidity, Wickedness and Waste*, ed. A. J. Marshall (William Heinemann, 1966), 75.

86 **Australian Reptile Park inventory**: Markwell and Cushing, *Snake-Bitten*, 126, 147.

86 **Casanova, a fifteen-foot freshwater crocodile**: Markwell and Cushing, *Snake-Bitten*, 97–100.

87 **"save one life a day"**: "Herpetologist Eric Worrell," video posted to YouTube by Cosmic Polymath, November 13, 2016.

88 **"hay, grains, fruit, mice"**: Markwell and Cushing, *Snake-Bitten*, 156.

88 **"It was a time of great crisis"**: Markwell and Cushing, *Snake-Bitten*, 157–67.

88 **a heart attack**: Markwell and Cushing, *Snake-Bitten*, 176.

CHAPTER 6: WELLS AND WELLINGTON

91 ***"Men renowned for their genius"***: Cervantes, *Don Quixote*, trans. John Rutherford (Penguin, 2000), 508.

98 **fell, breaking his neck**: "Rankin, Peter Robert," *Sydney Morning Herald*, January 4, 1979, 13; "Peter Rankin: A Biographical Sketch," Australian Museum, November 15, 2011.

98 **revised his book**: J. R. Kinghorn, *Snakes of Australia*, revised 2nd ed. (Angus and Robertson, 1964).

98 **the authoritative tome**: H. G. Cogger, *Reptiles and Amphibians of Australia* (A. H. Reed, 1975).

99 **more than two thousand different names**: H. G. Cogger et al., *Amphibia, Reptilia*, vol. 1 of *Zoological Catalogue of Australia* (Australian Government Publishing Service, 1983), 1–2.

101 **Cogger published his tome**: Cogger et al., *Amphibia, Reptilia*.

102 **the so-called bandit-hero**: Paul F. Angiolillo, *A Criminal as Hero: Angelo Duca* (Regents Press of Kansas, 1979), 2.

103 **their role as "great protector":** Angiolillo, *Criminal as Hero*, 43.

104 **articles by other authors:** Karen Kool, "Is the Musk of the Long-Necked Turtle, *Chelodina longicollis*, a Deterrent to Predators?," *Australian Journal of Herpetology* 1 (March 1981): 45–53; Garry A. Webb, "Observations on the Diving Ability of a Red-Bellied Black Snake," *Australian Journal of Herpetology* 1, no. 2 (March 1981): 54.

104 **The third issue:** Richard W. Wells and C. Ross Wellington, "A Synopsis of the Class Reptilia in Australia," *Australian Journal of Herpetology* 1, nos. 3–4 (December 31, 1983): 73–130.

105 **some 180 herpetologists:** R. Shine, "Report on the 1984 Australasian Herpetological Conference, and 1984 Annual General Meeting of the Australian Society of Herpetologists Held at Sydney and Springwood, 28 August to 2 September 1984," *Herpetological Review* 15, no. 4 (December 1984): 104–5.

106 **critics laid out their complaints:** Gordon C. Grigg and Richard Shine, "An Open Letter to All Herpetologists," *Herpetological Review* 16, no. 4 (December 1985): 96–97; Max King and Jeffrey Miller, letter to the editor, *Herpetological Review* 16, no. 1 (March 1985): 4–5; Harold Heatwole, letter to the editor, *Herpetological Review* 16, no. 1 (March 1985): 6; Michael J. Tyler, "Nomenclature of the Australian Herpetofauna: Anarchy Rules O.K.," *Herpetological Review* 16, no. 3 (September 1985): 69.

107 **"clearly have legal status":** Carl Gans, "Comment on Two Checklists," *Herpetological Review* 16, no. 1 (March 1985): 6–7.

107 **passed a series of motions:** Shine, "1984 Australasian Herpetological Conference."

107 **published a new edition:** Richard W. Wells and C. Ross Wellington, "A Classification of the Amphibia and Reptilia of Australia," *Australian Journal of Herpetology*, *Supplemental Series* 1 (March 1, 1985): 1–61; Wells and Wellington, "Synopsis of the Amphibia and Reptilia," 62–64.

108 **"W&W's taxonomy is unforgivably poor":** G. B. Monteith, "Terrorist Tactics in Taxonomy," *Australian Entomological Society News Bulletin* 21, no. 3 (August 1985): 66–69.

108 **"This may all seem to be very funny":** J. F. Veldkamp, "Terrorist Tactics in Taxonomy," *Flora Malesiana Bulletin* 9, no. 3 (1985): 311–12.

109 **"the world's biologists may find themselves beset":** Tony Thulborn, "Taxonomic Tangles from Australia," *Nature* 321 (May 1986): 13–14.

109 **Australian Society of Herpetologists' request:** President, Australian Society of Herpetologists, "Case 2531: Three Works by Richard W. Wells and C. Ross Wellington: Proposed Suppression for Nomenclatural Purposes," *Bulletin of Zoological Nomenclature* 44, no. 1–4 (March–December 1987): 116–21.

109 **eighty letters of support:** "Comments on the Proposed Suppression for Nomenclature of Three Works by R. W. Wells and C. R. Wellington," *Bulletin of Zoological Nomenclature* 45, no. 2 (June 1988): 153.

109 **as a group of French scientists pointed out:** Alain Dubois et al., "Comments on the Proposed Suppression for Nomenclature of Three Works by R. W. Wells and C. R. Wellington," *Bulletin of Zoological Nomenclature* 45, no. 2 (June 1988): 146–49.

109 **"We are outraged by this attitude":** P. Bouchet et al., "Further Comment on the Proposed Suppression for Nomenclature of Three Works by R. W. Wells and C. R. Wellington," *Bulletin of Zoological Nomenclature* 47, no. 2 (June 1990): 139–40.

109 **the commission delivered its decision:** International Commission on Zoological Nomenclature, "Three Works by Richard W. Wells and C. Ross Wellington: Proposed Suppression for Nomenclatural Purposes," *Bulletin of Zoological Nomenclature* 48, no. 4 (December 1991): 337–38.

115 **"substantiated by later, evidence-based studies":** David Williams et al., "The Good, the Bad, and the Ugly: Australian Snake Taxonomists and a History of the Taxonomy of Australia's Venomous Snakes," *Toxicon* 48 (July 14, 2006): 919–30.

115 **"vindicated Wells and Wellington":** Ken P. Aplin, " 'Amateur' Taxonomy in Australian Herpetology: Help or Hindrance?," *Monitor* 10, nos. 2–3 (1999): 104–9.

CHAPTER 7: WHISTLEBLOWER

117 *"I know their names but do not record them"*: Herodotus, *Histories*, 160.

119 **The first of his innumerable appearances:** "Hard Boiled Gamblers' Sweet Win," *Sydney Morning Herald*, January 18, 1973.

121 **signs of his future career:** Raymond Hoser, letter to the editor, *Herptile* 5, no. 3 (September 1980): 1–2.

121 **on Australian pythons:** Raymond Hoser, "Australian Pythons: Part One—Genera *Aspidites* and *Chrondropython*," *Herptile* 6, no. 3 (June 1981): 10–16.

121 **skin-shedding in captive Australian snakes:** Raymond Hoser, "Frequency of Shedding Skin in Captive Australian Snakes, Genera: *Liasis, Morelia,* and *Acanthophis*," *Herptile* 7, no. 3 (1982): 20–26.

121 **on the role of pelvic spurs:** Raymond Hoser, "The Role of Pelvic Spurs," *Herptile* 10, no. 3 (1985): 95.

121 **a short article about Hoser:** "Collector Must Send His Snakes to Zoo," *Sydney Morning Herald*, April 9, 1982.

121 **"the proud owner of 52 snakes":** Mark Coultan, "First There Was Monty Python, Now It's the Redfern Python," *Sydney Morning Herald*, May 21, 1983.

122 **raided his home, seized his snakes:** Andrew Keenan, "50 Snakes Seized During House Raid," *Sydney Morning Herald*, July 11, 1984.

122 **"the Snakeman of Redfern":** Keenan, "50 Snakes Seized."

122 **hundreds of dead snakes:** "Man Had 350 Reptiles in Home: Police," *Sydney Morning Herald*, July 13, 1984.

122 **concerned with the man's infatuation with snakes:** Keenan, "50 Snakes Seized."

123 **"feeding and copulating simultaneously":** Raymond T. Hoser, *Australian Reptiles and Frogs* (Pierson, 1989).

123 **an early proponent of whale conservation:** Raymond T. Hoser, *Endangered Animals of Australia* (Pierson, 1991).

124 **39 birds, half of which were dead:** "Bonanza in Birds for Aborigines," *Beverly (W. Aust.) Times*, September 14, 1972, http://nla.gov.au/nla.news-article201780949.

124 **possession of twelve wombats:** "Wildlife Smugglers Caught in SA," *Canberra Times*, March 31, 1971, http://nla.gov.au/nla.news-article110346110.

124 **"Mr. Bird-Brain":** Maurice Dunlevy, "Squad Probes 'Brain' in Bird Racket," *Sydney Morning Herald*, February 2, 1975.

124 **A Dutchman was intercepted:** Patrick Smithers, "Snakes Destined to Breed in Holland," *Melbourne Age*, January 18, 1985.

124 **sulphur-crested cockatoo:** "Bonanza in Birds for Aborigines."

124 **gang-gang cockatoos:** Leith Young, "Detailed Plans Aid Gangs That Pillage Eggs of Endangered Species," *Melbourne Age*, October 2, 1991.

124 **$20 a foot:** "Bird Smuggler Trail Leads to Melbourne," *Melbourne Age*, August 15, 1979.

124 **thieves broke into a wildlife sanctuary:** "Sneak Thieves Strip Sanctuary of Snakes," *Melbourne Age*, January 6, 1972.

125 **In a 1992 report:** Boronia Halstead, "Traffic in Flora and Fauna," *Trends and Issues in Crime and Criminal Justice* 41 (November 1992): 1–9.

125 **"well-known wildlife traffickers":** Raymond Hoser, *Smuggled*, rev. ed. (Kotabi Publishing, 1997), 1–5. Originally published by Apollo Books, 1993.

126 **"They were mixing it up with other meat, see":** Fia Cumming and Matt Condon, "Murder, Wildlife Smuggling Claims Rocks Parliament," *Sun Herald*, June 21, 1992.

126 **"highly defamatory":** "Removed Book Is Libellous, Says NPWS," *Sydney Morning Herald*, June 16, 1993.

126 **"very disturbing":** Jodie Brough, "Call for Inquiry on Fauna Trade," *Canberra Times*, June 21, 1993.

126 **cleared it of wrongdoing:** "ICAC Clears Wildlife Staff," *Canberra Times*, December 31, 1993.

126 **"tip of the iceberg":** "Removed Book Is Libellous."

127 **"tireless investigator":** H. Clare Callow, "The Strange Case of Raymond Hoser," *Australian Rationalist* 55 (2001): 29–33.

127 **dozens of key events:** Raymond Hoser, "The Fate of an Anti-Corruption Whistle-blower: Some Key Dates—an Abridged Chronology," smuggled.com.

128 **"instead of the museums where they belong":** Raymond Hoser, "Get Rid of Trams," *Melbourne Age*, January 4, 1990.

128 **"at no risk to anyone":** Raymond T. Hoser, "Police Cameras Are Putting Revenue Before Road Safety," *Melbourne Age*, April 16, 1993.

128 **"not a career bureaucrat":** Raymond Hoser, "Response to Taxi Death Inadequate," *Melbourne Age*, February 10, 1995.

129 **"nothing to do with Raymond Hoser":** Walter Connolly, "Taxi Driver Clarification," *Melbourne Age*, February 13, 1995.

129 **morning-show journalist Steve Price:** "3AW Transcript Steve Price," Mayne Report, December 6, 2010.

130 **forced to sell his house:** "By Crikey: The Article That Cost $25,000," *Sydney Morning Herald*, May 7, 2003.

131 **"He saw himself as a Robin Hood":** Keith Dunstan, *Saint Ned: The Story of the Near Sanctification of an Australian Outlaw* (Methuen Australia, 1980).

131 **"did not intend shooting the police":** "Ned Kelly: Meeting of Sympathizers," *Australian Town and Country Journal* (Sydney), November 13, 1880.

131 **"the myth remains the same":** Angiolillo, *Criminal as Hero*, 1.

132 **"Robin simply intervenes":** J. C. Holt, *Robin Hood* (Thames and Hudson, 1982), 10.

132 **a Joycean stream-of-consciousness:** "Ned Kelly's Jerilderie Letter: Transcript of the Jerilderie Letter Written by Bushranger Ned Kelly in 1879," National Museum Australia.

133 **"violent and vindictive man":** Russ Scott and Ian MacFarlane, "Ned Kelly—Stock Thief, Bank Robber, Murderer—Psychopath," *Psychiatry, Psychology and Law* 21, no. 5 (2014).

133 **killing "bloodily and with zest":** Holt, *Robin Hood*.

134 **feeding behavior of common scaly-foots:** Raymond T. Hoser, "Notes on the Feeding Behaviour of the Common Scaly Foot (*Pygopus lepidopodus*) and Burton's Legless Lizard (*Lialis burtonis*)," *Herptile* 10, no. 3 (1985): 93–94.

134 **immunity of snakes to their own venom:** Raymond T. Hoser, "On the Question of Immunity of Snakes," *Litteratura Serpentium* 5, no. 6 (1985): 219–32.

134 **snakes swallowing their own teeth:** Raymond T. Hoser and Chris Williams, "Snakes Swallowing Their Own Teeth," *Herpetofauna* 21, no. 2 (1991): 33.

134 **on hybridization among three species:** Raymond T. Hoser, "Problems of Python Classification and Hybrid Pythons," *Litteratura Serpentium* 8, no. 3 (1988): 134–39.

134 **sperm storage in the blue-tongue skink:** Raymond T. Hoser and Neil Simpson, "Sperm Storage in an Eastern Blue-Tongued Skink *Tiliqua scincoides* (Hunter, 1790)," *Bulletin of the Chicago Herpetological Society* 32, no. 4 (1997): 84.

134 **frequency of death adder shedding:** Raymond T. Hoser, "Frequency of Sloughing in Captive *Morelia*, *Liasis*, and *Acanthophis* (Serpentes)," *Herptile* 7, no. 3 (1982): 20–26.

134 **Mendelian genetics of death adders:** Raymond T. Hoser, "Genetic Composition of Death Adders (*Acanthophis antarcticus*; Serpentes; Elapidae) in the West Head Area," *Herptile* 10, no. 3 (1985): 96.

134 **sexual attractiveness of death adders:** Raymond T. Hoser, "Mating Behaviour in Australian Death Adders, Genus: *Acanthophis* (Serpentes; Elapidae)," *Herptile* 8, no. 1 (1983): 25–34.

134 **an inappropriately large lizard:** Raymond T. Hoser, "Note on an Unsuitable Food Item Taken by a Death Adder. (*Acanthophis antarcticus*) (Shaw)," *Herpetofauna* 13, no. 1 (1981): 30–31.

134 **"stuck my finger down his throat"**: Raymond T. Hoser, "Australia's Death Adders—Genus *Acanthophis*," *Reptilian* 3, no. 4 (1995): 7–21, and no. 5 (1995): 27–34.

135 **Ken Aplin holding a death adder:** Paul Bird, "Add an Adder," *Australian Post*, January 18, 1992.

135 **Ken Aplin was working on a formal description:** Simon Ball, "Further Data on the Pilbara Death Adder—a Suspected New Species," *Monitor* 5, no. 1 (1993): 5–10.

136 **It contained an article by Hoser:** "Death Adders (Genus *Acanthophis*): An Overview, Including Descriptions of Five New Species and One Subspecies," *Monitor* 9, no. 2 (April 1998): 20–41.

137 **head-to-tail physical description:** K. P. Aplin and S. C. Donnellan, "An Extended Description of the Pilbara Death Adder, *Acanthophis wellsi* Hoser (Serpentes: Elapidae), with Notes on the Desert Death Adder, *A. pyrrhus* Boulenger, and Identification of a Possible Hybrid Zone," *Records of the Western Australian Museum* 19 (1999): 277–98.

138 **he made his views clear:** Aplin, "'Amateur' Taxonomy."

139 **Hoser offered a response:** Raymond Hoser, "Herpetology in Australia—Some Comments," *Monitor* 10, nos. 2–3 (1999): 113–18.

CHAPTER 8: SNAKEMAN

141 *Danger, thrills and excitement*: Jim Copland, "Snake Man," *Sydney Morning Herald*, October 12, 1980, https://www.newspapers.com/image/123917054.

144 **the reason for his confidence:** Raymond Hoser, "Surgical Removal of Venom Glands in Australian Elapid Snakes: The Creation of Venomoids," *Herptile* 29, no. 1 (March 2004): 36–52.

145 **"These are the world's deadliest snakes":** "Snakeman Survives Bite from World's Deadliest Snake," video posted to YouTube by Snakeman Australia, September 28, 2008.

145 **"at no risk of being poisoned":** Marty Shlansky, "Slithering Snake Enthrals [*sic*] Crowd," *Standard* (Warrnambool, Vic.), July 8, 2009.

146 **amended its rules:** Raymond Hoser v. Department of Sustainability and Environment (Occupational and Business Regulation), Victorian Civil and Administrative Tribunal 2035, September 30, 2008, https://www8.austlii.edu.au/cgi-bin/viewdoc/au/cases/vic/VCAT/2008/2035.html.

146 **Hoser wrote to the department secretary:** Hoser v. Dept. of Sustainability and Environment VCAT 2035, 5.

146 **"Much of it was repetitive":** Hoser v. Dept. of Sustainability and Environment VCAT 2035, 7.

147 **"100% completely safe":** Mex Cooper, "'De-venomised' Snakes Ruled Dangerous," *Melbourne Age*, October 15, 2008, https://www.theage.com.au/national/devenomised-snakes-ruled-dangerous-20081015-5128.html.

150 **He called it *Pailsus pailsei*:** Raymond Hoser, "A New Snake from Queensland, Australia (Serpentes: Elapidae)," *Monitor* 10, no. 1 (1998): 5–9, 31.

150 **seven new python subspecies:** Raymond Hoser, "A Revision of the Australasian Pythons," *Ophidia Review* 1, no. 1 (October 2000): 7–27.

150 **he described a new species:** Raymond Hoser, "A New Species of Snake (Serpentes: Elapidae) from Irian Jaya," *Litteratura Serpentium* 20, no. 6 (December 2000): 178–86.

151 **"Taxonomic Contributions in the 'Amateur' Literature":** Wüster et al., "Taxonomic Contributions in the 'Amateur' Literature."

151 **"a sinister tale of vendetta":** Raymond Hoser, "*Pailsus*: A Story of Herpetology, Science, Politics, Pseudoscience, More Politics and Scientific Fraud," *Crocodilian* 2, no. 10 (September 2001): 18–31.

152 **the early days of the feud:** Raymond Hoser (rayhoser), posts under "Does this mean?," Kingsnake.com forum, November 21–22, 2002.

153 **new subspecies of taipan:** Raymond Hoser, "An Overview of the Taipans, Genus (*Oxyuranus*) (Serpentes: Elapidae) Including the Description of a New Subspecies," *Crocodilian* 3, no. 1 (May 2002): 43–50.

153 **two new subspecies of death adder:** Raymond Hoser, "Death Adders (Genus *Acanthophis*): An Updated Overview, Including Descriptions of 3 New Island Species and 2 New Australian Species," *Crocodilian* 4, no. 1 (September 2002): 5–11, 16–22, 24–30.

153 **four new species:** Raymond Hoser, "A New Subspecies of Elapid (Serpentes: Elapidae) from New Guinea," *Boydii* (Autumn 2003): 2–4; Hoser, "A Re-assessment of the Taxonomy of the Red-Bellied Black Snakes (Genus *Pseudechis*) with the Descriptions of Two New Subspecies," *Boydii* (Autumn 2003): 15–18; Hoser, "The Rough-Scaled Snakes, Genus: *Tropidechis* (Serpentes: Elapidae), Including the Description of a New Species from Far North Queensland, Australia," *Crocodilian* 4, no. 2 (August 2003): 11–14; Hoser, "A New Species of Elapid (Serpentes: Elapidae) from Western New South Wales," *Crocodilian* 4, no. 2 (August 2003): 19–26; Hoser, "Five New Australian Pythons," *Newsletter of the Macarthur Herpetological Society* 40 (August 2003): 4–9.

154 **including *Shireenhoserus*:** Raymond Hoser, "A Reclassification of the Pythoninae Including the Descriptions of Two New Genera, Two New Species and Nine New Subspecies," *Crocodilian* 4, no. 3 (November 2003): 31–37; no. 4 (June 2004).

154 **single new subspecies:** Raymond Hoser, "A New Subspecies of *Strophurus intermedius* (Squamata: Gekkonidae) from South Australia," *Boydii* (Spring 2005): 14–15.

154 **Prospero Productions:** Hoser v. Prospero Productions Pty Ltd & Ors, Federal Court of Australia 1376, October 8, 2004.

154 **letters to numerous newspapers and media:** "For the Record," *Melbourne Herald Sun*, October 8, 2010; Matt Meir, "Only One 'Snakeman,'" *Lismore (N.S.W.) Northern Star*, October 3, 2011; Kate Matthews, "Snake Man Strikes," *Grafton Daily Examiner*, April 2, 2012; Kate Matthews, "Snake Man Is Sprung," *Grafton Daily Examiner*, April 9, 2012.

154 **The phrase "snake man":** "New South Wales—No. VII," *Sydney Australian*, February 21, 1827, http://nla.gov.au/nla.news-article37071378.

154 **According to its website:** "Advantages," n.d., *Australasian Journal of Herpetology* website.

155 **detailed submission guidelines:** "Guidelines for Contributors," n.d., "Author Guidelines" link on *Australasian Journal of Herpetology* website.

155 **The first issue:** Raymond Hoser, "One or Two Mutations Doesn't Make a New Species: The Taxonomy of Copperheads (*Austrelaps*) (Serpentes: Elapidae)," *Australasian Journal of Herpetology* 1 (January 1, 2009): 1–28 (hereafter cited as *AJH*; all issues available on smuggled.com).

155 **The second issue:** Raymond Hoser, "Creationism and Contrived Science: A Review of Recent Python Systematics Papers and the Resolution of Issues of Taxonomy and Nomenclature," *AJH* 2 (February 3, 2009): 1–34.

156 **the journal's third issue:** Raymond Hoser, "A New Genus and a New Species of Skink from Victoria," *AJH* 3 (February 4, 2009): 1–6.

156 **In the fourth issue:** Raymond Hoser, "Eight New Taxa in the Genera *Pseudonaja* Gunther 1858, *Oxyuranus* Kinghorn 1923 and *Panacedechis* Wells and Wellington 1985 (Serpentes: Elapidae)," *AJH* 4 (February 9, 2009): 1–27.

156 **In the fifth issue:** Raymond Hoser, "Pain Makes Venomous Snakes Bite Humans," *AJH* 5 (February 10, 2009): 1–21.

156 **The sixth issue:** Raymond Hoser, "A Reclassification of the Rattlesnakes; Species Formerly Exclusively Referred to the Genus *Crotalus* and *Sistrusus*," *AJH* 6 (March 9, 2009): 1–21.

156 **Darwin scribbled a small tree:** Mark A. Ragan, "Trees and Networks Before and After Darwin," *Biology Direct* 4, no. 43 (November 16, 2009): 1–38.

157 **deepened the divide:** Yoon, *Naming Nature.*

157 **This comparative form:** I checked the truth of the Redditors' counterintuitive facts

using the "Divergence Time" tool on the Time Tree of Life website, which draws from published phylogenies to provide estimated divergence time between two taxa. For example, the last common ancestor of Spanish moss (*Tillandsia usneoides*) and oaks (*Quercus*) lived an estimated 160 million years ago, compared to that of oaks and spruces (*Picea*), which lived an estimated 330 million years ago.

CHAPTER 9: TAXONOMIST

159 **"attracts the half-scientists"**: Rolls, *They All Ran Wild*, 211.

159 **a short section on Hoser's taxonomic efforts**: Williams et al., "The Good, the Bad, and the Ugly."

161 **Hoser's taxonomy became personal**: Raymond Hoser, "A Reclassification of the True Cobras; Species Formerly Referred to the Genera *Naja*, *Boulengerina* and *Paranaja*," *AJH* 7 (March 23, 2009).

161 **scientists led by Christopher Kelly**: Christopher M. R. Kelly et al., "Phylogeny, Biogeography and Classification of the Snake Superfamily Elapidea: A Rapid Radiation in the Late Eocene," *Cladistics* 25, no. 1 (January 5, 2009): 38–63.

162 **led by Wüster himself**: Wolfgang Wüster et al., "The Phylogeny of Cobras Inferred from Mitochondrial DNA Sequences: Evolution of Venom Spitting and the Phylogeography of the African Spitting Cobras (Serpentes: Elapidae: *Naja nigricollis* Complex)," *Molecular Phylogenetics and Evolution* 45, no. 2 (November 2007): 437–53.

162 **Hoser's descriptions were invalid**: Van Wallach et al., "In Praise of Subgenera: Taxonomic Status of Cobras of the Genus *Naja* Laurenti (Serpentes: Elapidae)," *Zootaxa* 2236, no. 1 (September 9, 2009): 26–36.

163 **additional censures and charges**: Raymond Hoser v. Department of Sustainability and Environment, Victorian Civil and Administrative Tribunal 264, March 9, 2012.

163 **his whistleblowing exposé**: Raymond Hoser, "Sam the Scam: Sam the Koala is an Imposter!," *AJH* 8 (February 12, 2010): 1–64.

163 **court document included a transcript**: Hoser v. Dept. of Sustainability and Environment VCAT 264.

164 **"She'll take a bite for you now"**: "'I'd Do It Again': Snake Handler," *Yahoo News Australia*, August 9, 2011.

164 **"There's no reason"**: Anna Prytz, "Fangs for That, Dad—Handler Lets Reptiles Bite," *Manningham (Vic.) Leader*, August 10, 2011.

164 **"this is absolutely child abuse"**: Anna Prytz, "Snake Handler Backed," *Manningham Leader*, August 17, 2011.

164 **"use his children as guinea pigs"**: Prytz, "Snake Handler Backed."

164 **"Upon further reflection"**: "Showbiz as Usual for Snake Man," *Manningham Leader*, August 31, 2011.

164 **from Zimbabwe to Canada**: "Man Lets 'Deadly' Snakes Bite Daughter," *Zimbabwean*, December 8, 2011; Chuck Shepherd, "Man Takes Penguin-Obsession to the Extreme," *Shoal Lake Crossroads* (Man., Canada), October 28, 2011.

164 **Court documents say "several"**: Raymond Hoser v. Department of Environment, Land, Water and Planning, Victorian Civil and Administrative Tribunal B115/2012 & Z1060/2014, July 30, 2015.

164 **suspended his snake handler's license**: Hoser v. Dept. of Sustainability and Environment VCAT 264.

165 **published the ninth issue**: Raymond Hoser, "Exposing a Fraud! *Afronaja* Wallach, Wüster and Broadley 2009, Is a Junior Synonym of *Spracklandus* Hoser 2009!," *AJH* 9 (April 3, 2012): 1–64.

165 **his journal's tenth issue**: Raymond Hoser, *AJH* 10 (April 8, 2012): 1–63.

165 **In issue eleven**: Raymond Hoser, *AJH* 11 (April 8, 2012): 1–64.

165 **In issue twelve**: Raymond Hoser, *AJH* 12 (April 30, 2012): 2–76.

165 **paper after paper:** *AJH* 13 (June 30, 2012): 1–64; *AJH* 14 (June 30, 2012): 1–64; *AJH* 15 (July 6, 2012): 1–64.

166 **her dissertation:** Allyson Fenwick, "Beyond Building a Tree: Phylogeny of Pitvipers and Exploration of Evolutionary Patterns" (PhD diss., University of Central Florida, 2012).

166 **she included a phylogenetic tree:** Allyson M. Fenwick et al., "Morphological and Molecular Evidence for Phylogeny and Classification of South American Pitvipers, Genera *Bothrops*, *Bothriopsis*, and *Bothrocophias* (Serpentes: Viperidae)," *Zoological Journal of the Linnaean Society* 156 (2009): 617–40.

166 *Jackyhoserea*: Raymond Hoser, "A New Genus of Pitviper (Serpentes: Viperidae)," *AJH* 11 (April 8, 2012): 25–27.

166 *Allengreerus ronhoseri*: Raymond Hoser, "A New Genus and New Subspecies of Skink from Victoria," *AJH* 12 (April 30, 2012): 63–64.

166 **Australian Museum specimen number R129716:** Raymond Hoser, "Creationism and Contrived Science: A Review of Recent Python Systematics Papers and the Resolution of Issues of Taxonomy and Nomenclature," *AJH* 2 (February 3, 2009): 1–34.

166 **branch tip of its own:** Lesley H. Rawlings and Stephen C. Donnellan, "Phylogeographic Analysis of the Green Python, *Morelia viridis*, Reveals Cryptic Diversity," *Molecular Phylogenetics and Evolution* 27 (2003): 36–44.

167 **genera that Hoser would later rearrange:** R. Alexander Pyron et al., "The Phylogeny of Advanced Snakes (Colubroidea), with Discovery of a New Subfamily and Comparison of Support Methods for Likelihood Trees," *Molecular Phylogenetics and Evolution* 58 (2011): 329–42.

167 **"a new genus is erected":** Raymond Hoser, "A Three-Way Division of the African Centipede Eating Snakes, *Aparallactus* Smith, 1849 (Serpentes: Lamprophiidae: Aparallactinae) and a New Subgenus of Wolf Snakes *Lycophidion* Fitzinger, 1843 (Serpentes: Lamprophiidae, Lamprophiinae)," *AJH* 13 (June 30, 2012): 10–14.

167 *Nature* **published an article:** Brendan Borrell, "The Big Name Hunters," *Nature* 446 (March 15, 2007).

167 **"Hoser isn't a prolific scientist at all":** Jones, "Bad Scientists Are Threatening."

167 **"the Raymond Hoser problem":** Naish, "Taxonomic Vandalism."

167 **he'd been scooped by Hoser:** Elah Fader and Annie Minoff, hosts, *Undiscovered*, podcast, "Turtle v. Snake" episode, October 16, 2018.

168 **hadn't properly filed:** Wallach et al., "In Praise of Subgenera."

169 **his enemies' scheme:** Raymond Hoser, "A Review of the Extant Scolecophidians ('Blindsnakes') Including the Formal Naming and Diagnosis of New Tribes, Genera, Subgenera, Species and Subspecies for Divergent Taxa," *AJH* 15 (July 6, 2012): 1–64.

169 **forwarded him an email:** Hal Cogger confirmed to me in an email that he had alerted Raymond Hoser to the developing plan to boycott his names, writing that to keep the plan secret would have been "iniquitous."

169 **a stunningly direct attack:** Hinrich Kaiser et al., "Best Practices: In the 21st Century, Taxonomic Decisions in Herpetology Are Acceptable Only When Supported by a Body of Evidence and Published via Peer-Review," *Herpetological Review* 44, no. 1 (March 2013): 8–23.

170 **he appealed for help:** Raymond Hoser, "Case 3601 *Spracklandus* Hoser, 2009 (Reptilia, Serpentes, Elapidae): Request for Confirmation of the Availability of the Generic Name and for the Nomenclatural Validation of the Journal in Which It Was Published," *Bulletin of Zoological Nomenclature* 70, no. 4 (December 2013): 234–37.

CHAPTER 10: DAMNATIO MEMORIAE

171 *"has it heated your childish fancy"*: Friedrich Schiller, *The Robbers*, trans. Henry G. Bohn (London: George Bell and Sons, 1873), 75.

171 *"his name might be known to the whole world"*: "Valerius Maximus: Memorable

Deeds and Sayings," trans. Andrew Smith, Attalus, n.d., http://attalus.org/info/valerius.html.

172 **"Canyon Run at Mount Hotham"**: "Canyon Run at Mount Hotham with a Ski Crash Involving the Snakeman Raymond Hoser," video posted to YouTube by Snake Man, September 12, 2017.

172 **"Fastest Skier on Mountain"**: "Fastest Skier on Mountain—Fastest Snake Catcher—Snakeman Conquers Whistler BC," video posted to YouTube by Snakeman Australia, January 28, 2019.

172 **"There Is Only One Snakeman"**: "There Is Only One Snakeman in the USA, Canada and Australia," video posted to YouTube by Snakeman Australia, March 11, 2019.

174 **"The Applicant was a difficult witness"**: Hoser v. Dept. of Sustainability and Environment VCAT 264.

175 **The judges agreed:** Raymond Hoser v. Department of Sustainability and Environment, Supreme Court of Victoria Court of Appeal 206, September 5, 2014.

175 **not a "fit and proper person"**: Raymond Hoser v. Department of Environment, Land, Water and Planning, Victorian Civil and Administrative Tribunal B115/2012 & Z1060/2014, July 30, 2015.

175 **"argumentative and bombastic"**: Hoser v. Dept. of Environment, Land, Water and Planning VCAT B115/2012 & Z1060/2014.

178 **Articles 8.1.1.1 and 8.6:** George R. Zug, "Comment on the Proposed Confirmation of the Availability of *Spracklandus* Hoser, 2009 (Reptilia, Squamata, Elapidae)," *Bulletin of Zoological Nomenclature* 71, no. 4 (December 2014): 252–53.

178 **in violation of Article 9.7:** Wulf D. Schleip, "Comment on *Spracklandus* Hoser, 2009 (Reptilia, Serpentes, Elapidae): Request for Confirmation of the Availability of the Generic Name and for the Nomenclatural Validation of the Journal in Which It Was Published," *Bulletin of Zoological Nomenclature* 71, no. 1 (March 2014): 35–37.

179 **"potential problems with the fastening"**: Hinrich Kaiser, "Comment on *Spracklandus* Hoser, 2009 (Reptilia, Serpentes, Elapidae): Request for Confirmation [. . .]," *Bulletin of Zoological Nomenclature* 71, no. 1 (March 2014): 30–35.

179 **letters in Hoser's defense:** Thomas Cotton, "Comments on *Spracklandus* Hoser, 2009 (Reptilia, Serpentes, Elapidae): Request for Confirmation [. . .]," *Bulletin of Zoological Nomenclature* 71, no. 3 (September 2014): 181–82.

    The letter concludes with a note: "Comments supporting Case 3601 have also been received from Michael Smyth (private address, Melbourne, Australia), and Paul Woolf (President of the Herpetological Society of Queensland Incorporated). Those comments are noted and acknowledged, but are not published here because they repeat essentially the same arguments as those presented by Thomas Cotton (above)." Cotton and Smyth are both listed on the "Snakebusters Team" page on Raymond Hoser's smuggled.com website as Snakebusters staff.

179 **"most definitely published"**: Ross Wellington, "Comments on *Spracklandus* Hoser, 2009 (Reptilia, Serpentes, Elapidae): Request for Confirmation [. . .]," *Bulletin of Zoological Nomenclature* 71, no. 3 (September 2015): 222–26.

179 **"The issue is not 'Hoser' "**: Raymond Hoser, "Comment on *Spracklandus* Hoser, 2009 (Reptilia, Serpentes, Elapidae): Request for Confirmation [. . .]," *Bulletin of Zoological Nomenclature* 72, no. 1 (March 2015): 61–64.

179 **"Case 3601 represents a tipping point"**: Anders G. J. Rhodin et al., "Comment on *Spracklandus* Hoser, 2009 (Reptilia, Serpentes, Elapidae): Request for Confirmation [. . .]," *Bulletin of Zoological Nomenclature* 72, no. 1 (March 2015): 65–78.

180 **"The Commission is reluctant to suppress"**: International Commission on Zoological Nomenclature, "Opinion 2468 (Case 3601)—*Spracklandus* Hoser, 2009 (Reptilia, Serpentes, Elapidae) and *Australasian Journal of Herpetology* issues 1–24: Confirmation of Availability Declined; Appendix A (Code of Ethics): Not Adopted as a Formal

Criterion for Ruling on Cases," *Bulletin of Zoological Nomenclature* 78 (April 30, 2021): 42–45, http://dx.doi.org/10.21805/bzn.v78.a012.

181 **Hoser took it as an unqualified win:** Raymond Hoser, "Snakeman Raymond Hoser Wins ICZN Case 3601," smuggled.com, 2021.

181 **Acrantophiidae . . . *Macrochelys temmincki muscati*:** *AJH* 16 (April 29, 2013): 3–63.

182 **Ahaetulliini . . . Rhinocerophiina:** *AJH* 17 (April 29, 2013): 3–63.

182 *Klosevipera . . . Yunnanvipera:* *AJH* 19 (July 10, 2013): 3–63.

182 *Boulengerina adelynhoserae . . . Hawkeswoodelapidus:* *AJH* 20 (July 10, 2013): 3–63.

183 **Bennettsaurini . . . *Dasypeltis saeizadi*:** *AJH* 21 (October 20, 2013): 3–63.

183 *Funkiacrochordus . . . Pelochelys telstraorum:* *AJH* 22 (July 1, 2014): 3–64.

184 **Aspidomorphini . . . Borgsauriini:** *AJH* 23 (August 30, 2014): 3–64.

184 *Supremechelys . . . Snowdonsaurus:* *AJH* 24 (August 30, 2014): 3–64.

184 *Namibtyphlosaurus . . . Nessiini:* *AJH* 28 (July 1, 2015): 1–64; *AJH* 29 (July 1, 2015): 66–128.

185 *Aprasia parapulchella gibbonsi . . . Tympanocryptis bottomi:* *AJH* 30 (November 10, 2015): 3–64.

185 *Platyelapid . . . Denisonia gedyei:* *AJH* 31 (August 1, 2016): 3–63.

185 **Carphodactylini . . . *Lazarusus*:** *AJH* 32 (August 1, 2016): 3–63.

186 *Boiga irregularis halmaheraensis . . . Paramixophyes yeomansi:* *AJH* 33 (August 1, 2016): 3–64.

186 **Fiacumminggeckoini . . . *Lucasium rosssadlieri*:** *AJH* 34 (July 20, 2017): 1–63.

187 *Honlamopus . . . Wongaraneaus:* *AJH* 35 (July 20, 2017): 3–63.

187 *Aspidoclonion sloppi . . . Flamoscincus kintorei crossmani:* *AJH* 36 (March 30, 2018): 3–64.

187 *Gymnobelideus leadbeateri martinekae . . . Quattuorunguiscolotes grismeri:* *AJH* 37 (June 20, 2018): 3–64.

188 *Liopholis dannygoodwini . . . Scelotretus jenandersonae:* *AJH* 38 (August 10, 2018): 3–64.

188 *Paraheleioporus . . . Williamconnellysaurus:* *AJH* 39 (June 12, 2019): 3–63.

189 *Notanemoia . . . Caimanops amphiboluroides leucolateralis:* *AJH* 40 (July 10, 2019): 3–61.

189 *Greersaurus . . . Rankinia hoserae martinekae:* *AJH* 41 (August 1, 2019): 3–64.

190 *Burramys parvus hosersbogensis . . . Ctenophorus shireenhoserae:* *AJH* 42 (April 25, 2020): 3–64.

190 *Limnodynastes alexantenori . . . Simoselaps fukdat:* *AJH* 43 (April 25, 2020): 3–64.

190 *Adelynhoserhyleini . . . Maxinehoserranae brettbarnetti:* *AJH* 44 (June 5, 2020): 3–63.

190 *Maxinehoserranae maxinehoserae . . . Pustulatarana:* *AJH* 45 (June 5, 2020): 66–127.

192 *Pustulatarana longirostris tozerensis . . . Sandyranina:* *AJH* 46 (June 5, 2020): 130–91.

192 *Euprepiosaurus adelynhoserae . . . Assa roberteksteini:* *AJH* 47 (July 9, 2020): 3–63.

192 *Crotalus wellsi . . . Caudisona molossus smythi:* *AJH* 48 (August 3, 2020): 2–63.

192 *Ophioscincus paulwoolfi . . . Pogonomys sharonhoserae:* *AJH* 49 (August 6, 2020): 3–63.

193 *Bogophryne . . . Wellingtondellaini:* *AJH* 50–51 (October 10, 2020): 3–126.

193 *Chelydra haydnmcphiei . . . Lovelinaychelys:* *AJH* 52–53 (August 16, 2021): 3–126.

194 **Tetenditunguini . . . *Brunneisoculura*:** *AJH* 54 (October 14, 2021): 2–63.

195 **One of every ten:** Peter Uetz and Alexandrea Stylianou, "The Original Descriptions of Reptiles and Their Subspecies," *Zootaxa* 4375, no. 2 (January 24, 2018): 257–64.

CHAPTER 11: BABEL

197 ***Whatever once began, must have an end***: Rafinesque, *World, or Instability*, 12.

199 **"Routine, duty, responsibility"**: Angiolillo, *Criminal as Hero*, 3–4.

200 **their intention to continue the boycott**: Wüster et al., "Confronting Taxonomic Vandalism in Biology."

200 **"opiate of the unskilled and unsophisticated"**: L. Lee Grismer et al., "Phylogenetic Partitioning of the Third-Largest Vertebrate Genus in the World, *Crytodactylus* Gray, 1827 (Reptilia: Squamata: Gekkonidae) and Its Relevance to Taxonomy and Conservation," *Vertebrate Zoology* 71 (March 16, 2021): 101–54.

200 **"the result of these hesitations and uncertainties"**: Alain Dubois et al., "The Linz *Zoocode* Project. Second Report of Activities (2020). Nomenclatural Availability. 1. What Is Nomenclatural Availability?," *Bionomina* 28 (2022): 1–17.

201 **Geiger accused Chen, George, Averyanov**: Daniel L. Geiger, "Studies in *Oberonia*, 9: Lessons from Excess Names in *Oberonia* for Orchidaceae Systematics, Including a Revision of the *Oberonia* Sect. *Scytoxiphium*," *Lankesteriana* 21, no. 2 (2021): 139–56; D. L. Geiger, "Studies in *Oberonia* 8 (*Orchidaceae: Malaxideae*). Additional 24 New Synonyms, a Corrected Spelling, and Other Nomenclatural Matters," *Blumea* 65, no. 3 (2020): 188–203.

201 **vandals working on the true morels**: Michael Loizides et al., "Has Taxonomic Vandalism Gone Too Far? A Case Study, the Rise of the Pay-to-Publish Model and the Pitfalls of *Morchella* Systematics," *Mycological Progress* 21, no. 1 (2022): 7–38.

201 **decried the taxonomic vandalism**: Yasmina Marin-Felix and Andrew N. Miller, "Corrections to Recent Changes in the Taxonomy of the *Sordariales*," *Mycological Progress* 21, no. 69 (August 2022): 1–15.

201 **vandalizing the scarab beetles**: Matthew Moore et al., "Taxonomic Vandalism Is an Emerging Problem for Biodiversity Science: A Case Study in the Rutelini (Coleoptera: Scarabaeidae: Rutelinae)," abstract for presentation given at the Entomological Society of America, Portland, Ore., November 16–19, 2014.

201 **"work of apparent taxonomic vandalism"**: Kuo-Fang Chung et al., "An Updated Synopsis of *Lysimachia* L. (Lysimachieae, Primulaceae) of Taiwan," *Taiwania* 69, no. 2 (May 31, 2024): 207–28.

201 **"a digest of bad taxonomic practice"**: Philippe J. R. Kok, "Special Issue: A Few Steps Back, Several Steps Forward," *Journal of Vertebrate Biology* 72, no. 2303 (December 31, 2023): 1–4.

201 **"shared by all of our colleagues"**: Azad Teimori et al., "Comment on 'A Proposal for a New Generic Structure of the Killifish Family Aphaniidae, with the Description of *Aphaniops teimorii* (Teleostei: Cyprinodontiformes)' by Jörg Freyhof & Baran Yoğurtçuoğlu, Zootaxa (July 2020) 4810 (3): 421–451," *FishTaxa* 17 (2020): 15–16.

201 **"verging on taxonomic vandalism"**: Michael Balish et al., "Recommended Rejection of the Names *Malacoplasma* Gen. Nov., [. . .]," *Systematic and Evolutionary Microbiology* 69 (2019): 3650–53.

201 **committing taxonomic vandalism**: Barna Páll-Gergely et al., "Taxonomic Vandalism in Malacology: Comments on Molluscan Taxa Recently Described by N. N. Thach and Colleagues (2014–2019)," *Folia Malacologica* 28, no. 1 (March 2020): 35–76.

202 **"offensive words like 'vandalism'"**: Nguyen Ngoc Thach et al., "Comments on 'Vandalism' in Malacology," *Festivus* 52, no. 2 (2020): 184–90.

202 **the taxonomic vandalism wrought by Makhan**: The Editors, "Vandalism in Taxonomy," *Koleopterologische Rundschau* 77, no. 38 (July 2007).

202 **ignore the contents of Demangel Miranda's**: Jaime Troncoso-Palacios et al., "Without a Body of Evidence and Peer Review, Taxonomic Changes in Liolaemidae and Tropiduridae (Squamata) Must Be Rejected," *ZooKeys* 813 (January 7, 2019): 39–54.

202 **a genus of python that they called *Narawan***: Damien Esquerré et al., "Phylogenom-

ics, Biogeography, and Morphometrics Reveal Rapid Phenotypic Evolution in Pythons After Crossing Wallace's Line," *Systematic Biology* 69 (2020): 1039–51.

203 **the Wells and Wellington name should be used:** Hinrich Kaiser et al., "Nawaran [. . .] 2020 Is an Invalid Junior Synonym of Nyctophilopython Wells & Wellington, 1985 (Squamata, Pythonidae): Simple Priority Without Zoobank Pre-registration," *Bionomina* 20 (2020): 47–54.

203 **opening a "Pandora's box":** Dubois et al., "Linz *Zoocode* Project."

203 **"just another Hoser":** Stephen Thorpe, posts under "FW: Do rogue taxonomists need rogue publishers?," Taxacom mailing list, February 1, 2010.

203 **"A feeling of loneliness, alienation, mediocrity":** Albert Borowitz, *Terrorism for Self-Glorification: The Herostratus Syndrome* (Kent State University Press, 2005), xii, xiv–xv.

204 **"good good or bad bad":** Luisa Del Giudice, "Sabato Rodia's Towers in Watts: Art, Migration, and World Heritage," *Italian Canadiana* 31 (2017): 167–75.

204 **"Behold, the people is one":** Genesis 11:4–9 (King James Version).

205 **populations of nearly 5,500 species:** WWF, *Living Planet Report 2024: A System in Peril* (WWF, 2024).

205 **paradox of the coastline:** Benoit Mandelbrot, *The Fractal Geometry of Nature* (W. H. Freeman, 1982).

205 **bats:** Adriana Calahorra-Oliart et al., "Cryptic Species in *Glossophaga soricina* (Chiroptera: Phyllostomidae): Do Morphological Data Support Molecular Evidence?," *Journal of Mammalogy* 102, no. 1 (February 2021): 54–68.

205 **kiwi bird:** Maryann L. Burbidge et al., "Molecular and Other Biological Evidence Supports the Recognition of at Least Three Species of Brown Kiwi," *Conservation Genetics* 4 (March 2003): 167–77.

205 **cicadas:** Brian J. Stucky, "Morphology, Bioacoustics, and Ecology of *Tibicen neomexicensis* sp. n., a New Species of Cicada from the Sacramento Mountains in New Mexico, U.S.A. (Hemiptera, Cicadidae, *Tibicen*)," *ZooKeys* 337 (October 1, 2013): 49–71.

205 **butterflies:** Paul D. N. Hebert et al., "Ten Species in One: DNA Barcoding Reveals Cryptic Species in the Neotropical Skipper Butterfly *Astraptes fulgerator*," *PNAS* 101, no. 41 (October 12, 2004): 14812–17.

205 **pine trees:** Ann Willyard et al., "*Pinus ponderosa*: A Checkered Past Obscured Four Species," *American Journal of Botany* 104, no. 1 (2017): 161–81.

205 **"taxonomic inflation":** Nick J. B. Isaac et al., "Taxonomic Inflation: Its Influence on Macroecology and Conservation," *Trends in Ecology and Evolution* 19, no. 9 (September 2004): 464, 469.

206 **around two million:** Mark J. Costello et al., "Predicting Total Global Species Richness Using Rates of Species Description and Estimates of Taxonomic Effort," *Systematic Biology* 61, no. 5 (2011): 871–83.

206 **between two hundred million and nearly six billion species:** Brendan B. Larsen et al., "Inordinate Fondness Multiplied and Redistributed: The Number of Species on Earth and the New Pie of Life," *Quarterly Review of Biology* 92, no. 3 (September 2017): 229–65.

211 **describing new fish:** Raymond T. Hoser, "Two New Species of Fish, Previously Confused with the Macquarie Perch *Macquaria australasica* Cuvier 1830 (Actinopterygii: Perciformes: Percichthyidae) from East Coast Drainages in Australia," *AJH* 43 (April 25, 2020): 27–32.

211 **funnel-web spiders:** Raymond T. Hoser, "A New Genus-Level Classification of the Australian Funnel-Web Spiders (Hexathelidae: Atracinae)," *AJH* 35 (July 20, 2017): 54–56.

211 **a tree kangaroo:** Raymond T. Hoser, "A New Species of Tree Kangaroo, Genus *Dendrolagus* Müller, 1840 from Tembagapura, Mimika, Irian Jaya, Indonesia," *AJH* 40 (July 10, 2019): 50–55.

211 **rock wallaby, pygmy possums, ring-tailed possums, yellow-bellied gliders, and potoroos:** *AJH* 42 (April 25, 2020): 3–49.

211 **mammalian taxonomic vandalism:** S. M. Jackson et al., "The Importance of Appro-

priate Taxonomy in Australian Mammalogy," *Australian Mammalogy* 45, no. 1 (October 12, 2022): 13–23.

211 **"a killing machine":** Claudia Poposki, "Expert Snake Catcher Says Coronavirus Has Made This Year's Serpent Season Start Early—and Reveals Why You're More at Risk of a Deadly Encounter Due to Lockdown," *Daily Mail Australia*, August 7, 2020.

211 **"They are deadly":** Freya Noble, "Deadly Tiger Snake Found Wrapped Around Petrol Pump in Suburban Melbourne," *9News*, September 16, 2020.

211 **"appears to be merely a stunt":** "Snake Catcher 'Staged' Deadly Encounter on Melbourne Petrol Pump," *9News*, September 17, 2020.

211 **"I don't plant snakes to generate business":** Charlotte Karp, "Snake Catcher Furiously Denies Planting a Deadly Tiger Snake at a Melbourne Petrol Pump After Coles Called the Police," *Daily Mail Australia*, September 19, 2020.

211 **arguing in his filing:** Raymond Terrance Hoser v. Australian Broadcasting Corporation and Paul James Berry and Misa Han, County Court of Victoria at Melbourne 1164, October 6, 2022.

212 **"faker and a fraud":** Charlotte Grieve and Kishor Napier-Raman, "Slippery Suit," *Melbourne Age*, October 11, 2022.

212 **public use of the phrase "snake man":** Raymond Hoser v. Euroa Agricultural Society Inc and ORS, Federal Circuit and Family Court of Australia MLG2279, 2022.

212 **his most ambitious legal gambit yet:** Raymond Hoser v. Arthur Georges and ORS, Federal Circuit and Family Court MLG2313, 2022.

212 **each pay him $200,000:** Hoser v. Arthur Georges and ORS, MLG2313, 2022; notice of filing and hearing, 8/11/2022, 9:30 a.m.

213 **"We . . . completely disregard Hoser's nomenclature":** Alex Slavenko et al., "Revision of the Montane New Guinean Skink Genus *Lobulia* (Squamata: Scincidae), with the Description of Four New Genera and Nine New Species," *Zoological Journal of the Linnaean Society* 195, no. 1 (2022): 241–43.

213 **replaced more than one hundred of his names:** Wüster et al., "Confronting Taxonomic Vandalism in Biology."

213 **"censuring taxonomic vandals":** Aurélien Miralles et al., "Species Delimitation Methods Put into Taxonomic Practice: Two New *Madascincus* Species Formerly Allocated to Historical Species Names (Squamata, Scincidae)," *Zoosystematics and Evolution* 92, no. 2 (December 6, 2016): 257–75.

214 **"A man naming one Shireen":** Amanda Joy, "Medusa and the Taxonomic Vandal," in *Snake Like Charms* (UWA Publishing, 2017), 47.

215 *Andrias another . . . Scienceneedsevidence toprogress***:** Raymond T. Hoser, *AJH* 80 (August 1, 2025): 3–63.

215 *Aa aaaaaugh . . . S. murderingpoliceorum***:** Raymond T. Hoser, *AJH* 76–78 (May 7, 2025): 1–192.

215 *H. nswpolicearecrooks . . . Scored yetanother***:** Raymond T. Hoser, *AJH* 75 (May 6, 2025): 1–64.

215 *S. exy***:** Hoser, *AJH* 76–78.

215 *O. arukidding . . . C. wereisdat***:** Raymond T. Hoser, *AJH* 71 (June 12, 2024): 1–64.

215 *Oh kay . . . Oh phuk***:** Raymond T. Hoser, AJH issue 70, May 27, 2024): 1–64. Raymond T. Hoser, *AJH* 68 – 69 (May 20, 2024): 1–128.

215 *C. whatdafuk, C. fukdat***:** Hoser, AJH 71. Hoser, *AJH* 76–78 AJH 76–78.

215 *F. tasteslikesheet***:** Hoser, *AJH* 71.

215 *S. ivebeenshaton***:** Hoser, *AJH* 76–78.

215 **"hard-core criminals":** Raymond T. Hoser, *AJH* 79 (June 16, 2025): 1–64.

215 **"self-professed pseudo-scientists":** Hoser, *AJH* 80.

215 **"sex offenders and thieves":** Hoser, *AJH* 79.

215 **terrorism:** Hoser, *AJH* 79.

215 **"Gulf of Wolfgang Wüster":** Hoser, *AJH* 80.

INDEX